Step-by-Step QFD
Customer-Driven Product Design

Second Edition

John Terninko

St. Lucie Press
Boca Raton, Florida

D1127780

Library of Congress Cataloging-in-Publication Data

Catalog record is available from the Library of Congress.

Visit our website at www.crcpress.com.

© 1997 by CRC Press LLC
St. Lucie Press is an imprint of CRC Press LLC
No claim to original U.S. Government works
International Standard Book Number 1-57444-110-8
Printed in the United States of America   3 4 5 6 7 8 9 0
Printed on acid-free paper

TABLE OF CONTENTS

Foreword

In the 1960s, Quality Control and Quality Improvement had a distinctively manufacturing flavor in Japan. New products were introduced to manufacturing with many start-up problems. Teams of workers, managers and engineers went to work to eliminate the problems and eventually they produced product of a high level of quality. This lasted until the next new product was introduced and the cycle started again.

Finally, in the late 1960s and early 1970s, Joji Akao and others went to work on improving the design process so that when the new product was introduced to manufacturing, it was high quality from the beginning. The process for improving design was called Quality Function Deployment (QFD). From 1975 to 1995, this tool/process was integrated with other improvement tools to generate a mosaic of opportunities for product developers.

The QFD mosaic as a whole is impressive and foreboding. John Terninko, one of the first to study QFD in the US, has dismantled and demystified the mosaic so that each piece can be studied and mastered independently. By following the blueprint of this very readable book, it becomes possible to learn each piece of QFD, step by step, and to learn to apply the principles and tools over multiple developmental projects.

Only deep thought and much work make it possible to take complex things and make them simple. John has accomplished this task wonderfully well in this step-by-step approach to QFD. I hope you find it as enjoyable and rewarding a journey as I did.

Bob King
GOAL/QPC

Preface

After many years of consulting and training, I have finally listened to the customer, the beginning practioner. You want a world-class new product design. You want to tailor your design effort to your unique needs. This manual is a step-by-step presentation of Quality Function Deployment (QFD) activities necessary for implementing a successful customer-driven product design.

Many organizations experience significant breakthroughs early in the QFD process. However, I believe you are interested in a world-class, customer-driven design rather than in using QFD. Systematic Innovation has been added to the second edition to facilitate the design of world-class products. Systematic Innovation is also known as TRIZ (the Russian acronym for the *Theory of Inventive Problem Solving*).

For many, QFD is a new way of thinking which takes time to absorb. The format used here allows you to learn comfortably over an extended period of time. You can leave off and pick up at a later date when ready for the next level. Take as many steps as you find necessary, at the rate that works best for you.

See you at the end of the manual.

ACKNOWLEDGMENTS

For almost everyone who writes a book or training manual, ideas come from many sources. The following sources are the ones that come to mind, and acknowledging them in no way implies that I did not receive help from others, particularly seminar participants.

The heated hours of discussion at the Quality Function Deployment Institute (QFDI) Forum meetings have always provided food for thought.

Discussion and problem solving with Bob King and John McHugh during the formative years of QFD in the USA expanded my vision.

Also, my participation on the GOAL/QPC research committee provided an excellent forum for discussion.

Satoshi "Cha" Nakui introduced a comprehensive QFD model with the Voice of the Customer Table (VOCT).

Stan Marsh of GOAL/QPC always asked the fundamental questions that kept me on track.

Glenn Mazur, the primary translator of Japanese quality material, has been an invaluable colleague. Some of his earlier writing forms the backbone of the brief QFD history in Chapter 1.

Alberto Gomez Tellechea, came to study with me for a year and taught me.

Richard Zultner, who must see a theory before accepting it, has kept me honest in my explanation. Richard is always pushing the models and integrating disciplines. Special thanks are due to him for the graphic of coherent designs and QFD breadth and depth in Chapter 1.

Thanks to Brian Guercio for providing the cartoon of the mountaineers.

Many of my clients have forced me to stay in the real world. Mary Sonnack and Tom Swails (of 3M) created the easel pad case study used throughout this book.

Boris Zlotin and Alla Zusman of Ideation International graciously contributed their time and insights while I worked to understand TRIZ.

My technical writer, Maggie Rogers, took all the scraps of paper and our discussions of what I really wanted to say and somehow put them together to make them understandable. When she started, Maggie had never heard of QFD. Now she is considering printing her own QFD business cards after she becomes a mom.

Special thanks to Candy, who allowed me to have a closer relationship with the computer as the deadline approached. She was always there, pulling me back to reality.

Thanks to Mary Ann Kahl for the final touches.

Errors or omissions are entirely my own.

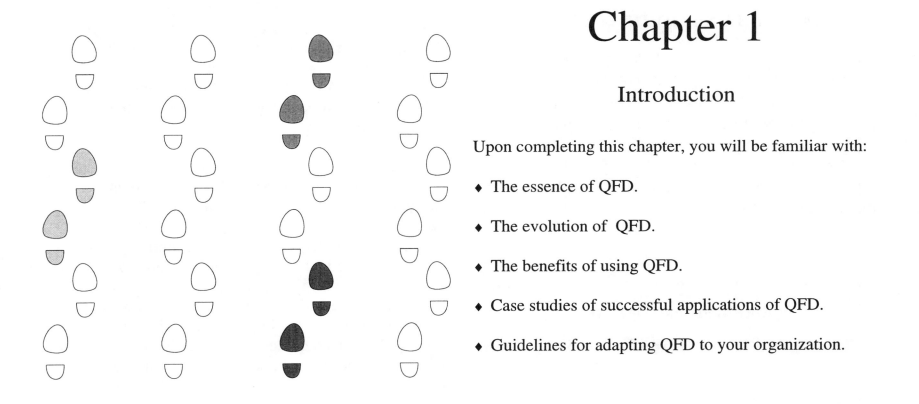

Chapter 1

Introduction

Upon completing this chapter, you will be familiar with:

♦ The essence of QFD.

♦ The evolution of QFD.

♦ The benefits of using QFD.

♦ Case studies of successful applications of QFD.

♦ Guidelines for adapting QFD to your organization.

Introduction

This manual will be your guide for the journey through the design process, beginning with the murmurs of your customer and ending with increased market share. The book is primarily intended for three types of readers: people who are studying Quality Function Deployment (QFD) for the first time, practitioners in their first year of using QFD and organizations that are experiencing manufacturing difficulties. While the manual includes information for model upgrades, it focuses on new product design.

Included in this manual are a number of features that enable readers to learn about QFD on their own, independent of a QFD course or seminar. The workshops, appendix and glossary function as guides, helping QFD students complete their journey toward a world-class product. Many journeys in our lives include resting places where we can stop while climbing a mountain and reap the benefits of the sights without going all the way to the highest peak. Similarly, *Step-by-Step QFD* points out intermediate destinations where the QFD practitioner may choose to pause and implement one aspect of QFD. Some readers may choose to terminate the QFD process after using only one tool; others may decide to complete the book and embark upon a more comprehensive QFD application.

QUALITY FUNCTION DEPLOYMENT

QFD IS A DETAILED SYSTEM FOR TRANSLATING THE NEEDS AND WISHES OF THE CONSUMER INTO DESIGN REQUIREMENTS FOR PRODUCTS OR SERVICES.

DETAILED ANALYSIS CAN BE EXTENDED TO THE DESIGN OF SYSTEMS, PARTS, PROCESSES AND CONTROL MECHANISMS.

THIS RESULTS IN

GREATER PROFITS AND

INCREASED MARKET SHARE.

Figure 1-1

Before starting the journey, it is helpful to define QFD and to examine recent QFD projects. QFD is a modern quality system aimed at increasing market share by satisfying the customer. This system strategically selects and makes visible customer requirements that are important for outperforming the competition.

Once customer needs are understood, they are translated into design requirements. The subjective desires of the customer are mapped into the language of the engineer. QFD focuses on delivering value by understanding the customer's wants and needs and then deploying these expectations throughout the development process. This includes identifying the best values for components, parts or materials. To manufacture a world-class product, the customer information must be deployed to the manufacturing process and control systems. A complete design process would include packaging, delivery and support. The end results are increased market share and greater profits.

These benefits are the result of reduced development time by a factor of 2 or 3. Reducing the number of engineering changes can significantly increase profits. Companies frequently set aside large sums of money as contingency for introduction problems. With QFD, introduction costs decrease and market share grows because of satisfied customers who have a product that meets or exceeds their needs and desires. Besides reducing production costs, QFD facilitates communication and cooperation among various functions of an organization.

QUALITY FUNCTION DEPLOYMENT

SHORTER DEVELOPMENT TIME (1/2 TO 1/3)

FEWER ENGINEERING CHANGES (1/2 TO 1/4)

REDUCED INTRODUCTION COSTS

SATISFACTION OF CONSUMER NEEDS AND DESIRES

IMPROVED PRODUCT MANUFACTURABILITY

COMMONALITY OF LANGUAGE

DEVELOPMENT OF A READY REFERENCE
FOR THE FUTURE

Figure 1-2

Traditional Design Process

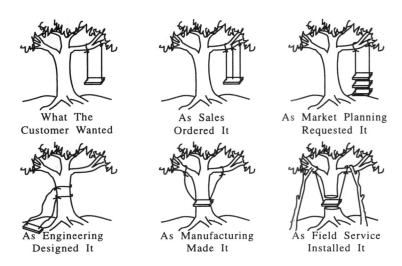

| What The Customer Wanted | As Sales Ordered It | As Market Planning Requested It |
| As Engineering Designed It | As Manufacturing Made It | As Field Service Installed It |

QFD Design Process

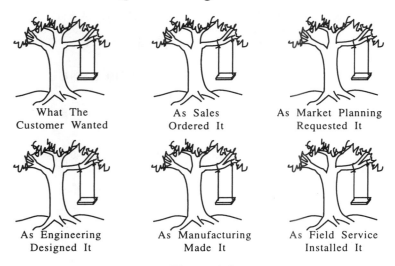

| What The Customer Wanted | As Sales Ordered It | As Market Planning Requested It |
| As Engineering Designed It | As Manufacturing Made It | As Field Service Installed It |

Figure 1-3

Another substantial benefit of the QFD process is the development of a common language. Discussions during the entire design process identify miscommunications resulting from contrasting value systems or different vocabularies. An active glossary of terms is developed and linked to each analysis activity. Teams also produce matrices and tables that provide further documentation of their activities. These documents become increasingly comprehensive, providing a ready reference for related products and future model upgrades. With its creation of a common language and thorough documentation, QFD develops an organization's sensitivity to the customer's demands. QFD designs value into your product. Only a focused effort can make a product so good that customers boast about it—the true test of the delighted customer.

QFD builds quality into products by fostering a coherent design process. Consumers generally make subjective statements when evaluating a product. QFD employs a variety of strategies to systematically translate this feedback into objective design requirements. These design requirements are communicated to all relevant organizational functions so that everyone is working toward a synergistic whole.

As seen in an old *Readers Digest* cartoon (Figure 1-3), the customer, a child, wants a simple rope swing. The sales person decides to enhance the design with three ropes. Marketing envisions three happy children in its sales campaign. The engineering staff has become sensitive to injured party lawsuits and does not want anyone to fall. The manufacturing team believes it knows what the customer really wants. Finally, because of a lack of training, the field person does not know how to install the swing. In this hypothetical example, every organizational function has its own version of what the customer desires. Each version reflects a different group's unique view of the world. The QFD process aligns the entire organization.

Few experiences are more discouraging than doing your best, only to discover over the course of a project that your labors will not make a difference. Figure 1-4 further illustrates the incoherent process that frustrates talented employees. The upper figure shows a concept team that is proud of three aspects of its design, as indicated by the X's. The team passes these concepts on to functional unit 1, possibly the development team. This second group values a different set of strengths, but the two that they are most proud of do not align with the concept team's three contributions. This misalignment of effort can continue throughout the development process. If none of the X's are in line with the most important customer need, this need becomes lost in the design process. This approach results in a game of chance, indicated by the large question mark.

The bottom portion of Figure 1-4 illustrates how QFD aligns the entire organization with the customer's perspective. By concentrating everyone's efforts on the customer's needs, QFD involves all of the organization's functions in the design process. This focus promotes a coherent process where practitioners can focus on what they do best and know that their efforts count.

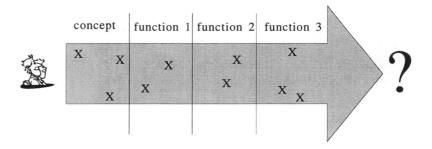

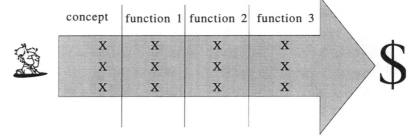

Adapted from R.E. Zultner "Task Deployment for Service: Process QFD" Transactions for the Fourth Symposium on Quality Function Deployment, 1992

Figure 1-4

The Role of the Customer in Design

Traditional marketing techniques simply do not provide the level of detail utilized by the normal QFD process. QFD tools examine the customer needs in detail from a variety of angles. Some marketing organizations have built their reputations upon these precise methods. At the 1990 Novi Symposium, Robert Klein of Applied Marketing Science, Inc. presented the integration of several marketing improvements into a process he called VOCALYST^{TM}. The study originated from John R. Hauser's research at MIT's Sloan School of Management using a four-phase process: interviews with customers, transcription of the tape-recorded interviews, data grouped in a variation of the affinity diagram and cluster analysis to calculate the importance of customer needs.

With its intensive study of customer needs, QFD has generated innovative strategies for outperforming the competition. For instance, the Kano Model reveals that zero defects for basic needs does not mean a satisfied customer. Does the product do what is desired in a way that is exciting? The features which result in increased satisfaction with improvements in performance are the common indicators for a happy customer. The uncommon feature will catch the competition off stride. An organization must identify some excitement features for its product. Some features are not expected. Just a hint of the feature excites and makes the customer happy. In high tech industries, today's exciting features are tomorrow's expected features. In order to ensure the production of quality products, organizations must keep in touch with their customers so that they know how and when their needs are changing. Chapter 4 addresses the Kano Model in greater detail.

A playful definition of quality underscores the importance of this intimacy between an organization and its customers:

"Quality is what makes it possible for a customer to have a love affair with your product or service. Telling lies, decreasing the price or adding features can create a temporary infatuation. It takes quality to sustain a love affair.

Love is always fickle. Therefore, it is necessary to remain close to the person whose loyalty you wish to retain. You must be ever on the alert to understand what pleases the customer, for only customers define what constitutes quality. The wooing of the customer is never done." (Myron Tribus in ASQC Statistics Division Newsletter, 1990)

To really understand and "woo" its customers, an organization must go to their environment and actually observe the customers using the product. The Japanese call this form of market research *"going to the Gemba."* This journey leads to a better understanding of the product's context of application, a vital component of product design. Figure 1-5 provides a concrete example of how this context of application shapes designs. In this example, an organization designs privately owned transportation vehicles. One customer states seven requirements, starting with "good fuel economy." The old-style VW was an exciting solution in the 1950s-1970s, but this is no longer the case. Another customer states the same requirements, but the most appropriate design looks nothing like the first design. The second customer lives in a location without paved roads or gasoline stations. There are heavy rains and no bridges. Going to the *Gemba* and viewing the context that shapes customer requirements is an essential step for creating appropriate designs.

The following case studies of American applications of QFD demonstrate the value of going to the *Gemba,* thus affirming Tribus's assertion that "The wooing of the customer is never done."

Demanded Quality for Transportation

Good Fuel Economy
Good Road Stability
Good Ride
Carries Heavy Load
Sporty Style
Low Cost
Uses Multigrade Fuel

Demanded Quality for Transportation

Good Fuel Economy
Good Road Stability
Good Ride
Carries Heavy Load
Sporty Style
Low Cost
Uses Multigrade Fuel

Figure 1-5

Going to the *Gemba* and a Successful QFD Project

When looking at organizations using QFD, it is important to consider the *breadth* and *depth* of their applications. A company using an application that has depth will employ QFD activities for a variety of different processes; these might include customer research, design activities and the manufacturing processes. A QFD project that has *depth* contains a variety of analyses. The *breadth* of application is a measure of QFD's persuasiveness throughout an organization. QFD consultant Richard Zultner coined these phrases. The following three case studies illustrate QFD applications, each one involving more depth than its predecessor. None of the case studies indicate the breadth of the organization's application. Implementations of QFD containing breadth occur in mature companies that have been using QFD and quality control for a number of years. Each of these cases is explained in greater detail in Appendix A.

It may not be fair to call an effort a QFD project if the organization stops the process after recognizing that it has overlooked an important customer. However, initiation of the QFD process by Rehab Concepts of Willington, MA, helped the company identify a customer voice it had not previously "heard." Rehab Concepts offers physical therapy for employees who have experienced industrial accidents. The QFD team believed the most important customer need was that of the employer and the insurance company for the employee to return to work as soon as possible. Interviewing patients for one hour revealed that these customers' primary need was to be able "to do everything they did before," such as skiing and playing tennis. The company's previous analyses were driven by its source of income, the insurance company and the employer. Today, with this new understanding of who it is servicing, Rehab Concepts has an excellent reputation and its services are considered second to none in the state of Massachusetts.[1]

One Matrix and the *Gemba* Recapture a Lost Market

The case of Puritan-Bennett resembles Rehab Concepts in that only a few QFD tools helped the company reap impressive benefits. After researching customers' needs, Puritan-Bennett required only one matrix to show how to best improve its product.

Puritan-Bennett designs and manufactures spirometers. Hospitals and clinics use these devices to measure lung function and detect pulmonary disease. In 1988, the product sold for $4500 US. Then a new competitor offered a product with fewer features that sold for $1995 US. This was less than half the current cost to Puritan-Bennett. Using the QFD process, Puritan-Bennett was able to develop a new model with more features for $1590 US.[2]

A QFD Application with Depth

Initiating the QFD process at Kimberly-Clark produced 16 different matrices in a comprehensive study which resulted in major breakthroughs. Kimberly-Clark successfully applied QFD in the following areas:

1. Creating a new product design.
2. Using a new manufacturing process.
3. Using a new technology.
4. Manufacturing in a new plant.
5. Manufacturing with new equipment (one of a kind and some first of a kind).
6. Working with new employees (Scheurell 1992, 1993).[3]

Preparing for QFD

Implementing QFD for new product designs requires a substantial initial investment of resources in the form of time, money and staffing. In contrast, the traditional allocation of resources begins very modestly and increases to a peak. The peak occurs after major problems have been identified and the customers have been waiting for corrective action. Of course, convincing the financial manager to front-load the project can be difficult. But if you do not, you may experience a situation similar to the scenario Rehab Concepts underwent: you will have lost six months of design time.

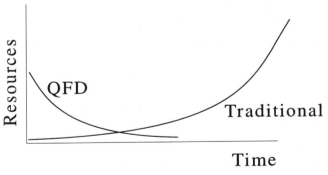

Allocation of Resources

Figure 1-6

1 Thanks are due to Jim Bruer for allowing me to share the Rehab Concepts case.

2 The Puritan-Bennett case is shared thanks to Robert Klein of Applied Marketing Science, Inc.

3 Thanks to Diane Scheurell of Kimberly-Clark for sharing the details of this case.

QFD and Increased Efficiency

The reduction in startup costs that Kimberly-Clark experienced underscores the difference between QFD and traditional production methods. A QFD-designed car has fewer engineering changes and adjustments earlier in the development cycle. The number of engineering changes/units of time reaches a peak about 19 months before the first day of production (Figure 1-7). There are virtually no changes after the first day of production. Organizations using QFD expect the prototype to perform correctly, for its success is a confirmation of their thoughtful design.

In contrast, many Western corporations set aside large sums of money for startup problems. If something less than or close to the anticipated startup cost is experienced, everyone is happy. They believe that their designs are so inadequate that they budget for quality problems after the first day of production. Only if the budget is exceeded does anyone become concerned with the cost of customer complaints.

Accordingly, the traditional approach has a growing number of engineering changes until shortly before the product goes on the market. Once production starts, the decline proves to have been only temporary as the customers discover errors in design. A second peak occurs shortly after the first day of production. The cause of the first peak is the process of design. This curve represents an organization that designs, builds,

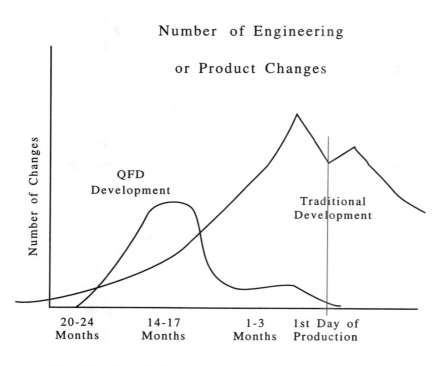

Number of Engineering

or Product Changes

Adapted from L.P. Sullivan, "Quality Function Deployment," Quality Progress, June 1996

Figure 1-7

tests and redesigns a prototype. The organization expects that the prototype will fail. The workers repeat this process until the design is acceptable or they have run out of time. This method involves essentially 100% inspection of the design process. One hundred percent inspection never works for products that do not have quality designed into them, nor is it a solution for a poor design process.

Yoji Akao says the Japanese actually use both curves. The curve identified by a peak 19 months before the first day of production represents the effort devoted to solving important problems early in the design process. The other curve represents the effort devoted to relatively unimportant items, such as the color of a car's seats.

During the mid '80s at a public QFD seminar in Detroit, some American car manufacturers discussed the number of engineering changes made to two successful US cars. One car had 1,791 new tooled parts with 5,987 engineering changes. This new model had an average of 3.3 engineering changes per new part. Another car, with 2,700 new tooled parts, also had an average of 3.3 engineering changes per new part. When the car manufacturers shared these figures, a participant from the computer industry said that these averages matched his organization's figures. Redesigning a product 3.3 times wastes money and time. Think of all the resources the company could have saved if it had started with a careful analysis of its design process.

Recent history shows that auto manufacturers can increase their market share if they base design processes on their customers' demands. Chrysler Corporation used QFD to design the 1994 Neon. The following was presented at the 1994 annual QFD symposium, but it has not been shared with the general public: The team concentrated on the customer need Fun to Drive. Market reports during the fall of 1995 continued to indicate that the Neon is a successful entry level car. Here then is dramatic proof of QFD's effectiveness.

> ## "Measure a thousand times and cut once."
>
> Turkish Proverb

Old-fashioned common sense and discipline form the bedrock upon which QFD's achievements rise.

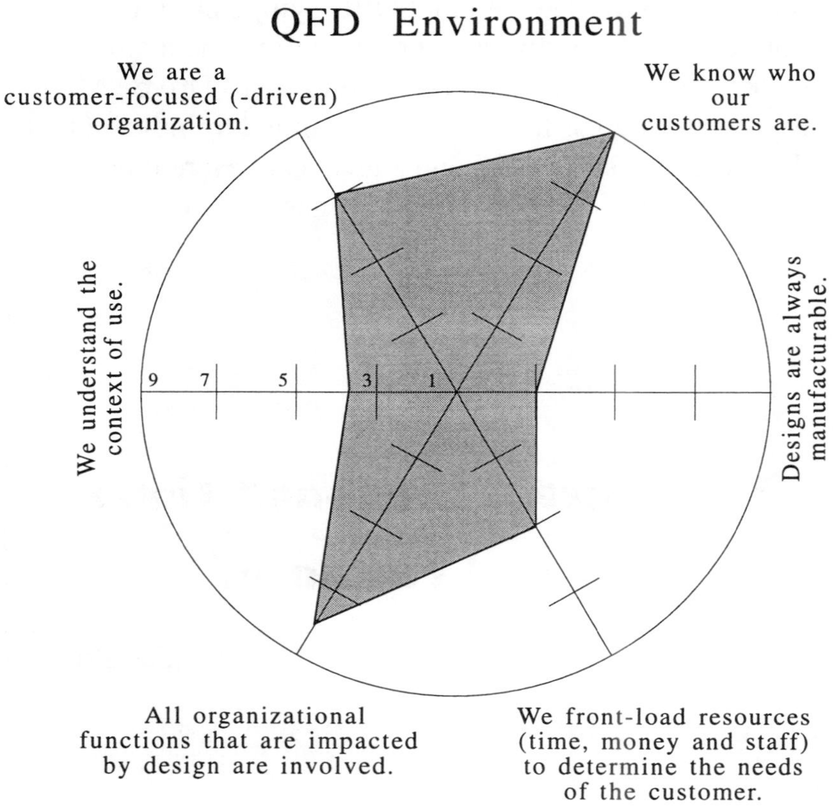

QFD Environment

We are a
customer-focused (-driven)
organization.

We know who
our
customers are.

We understand the
context of use.

Designs are always
manufacturable.

9 7 5 3 1

All organizational
functions that are impacted
by design are involved.

We front-load resources
(time, money and staff)
to determine the needs
of the customer.

Figure 1-8

A radar chart provides a graphic way to examine your organization's QFD environment. The chart is a guideline for discussion. Results are dependent upon the context and character of your organization. The QFD process in general, assumes that all teams are cross-functional. The chart uses a nine- point scale to record how well each statement describes your organization. A "1" in the center of the circle indicates that the statement is the opposite of your organization. A "9" on the perimeter indicates that the statement describes your organization precisely.

The six dimensions in the radar chart are frequent issues for QFD teams. For the upper left-hand statement, "We are a customer-focused (-driven) organization," this team selected a 7 to indicate that this is often the case for their organization. A customer-focused design may be appropriate for a technology-driven organization, although a customer-driven design may be too limiting for a technology-driven product. Which is best for you?

Each of the other five areas on the chart has been discussed in this introduction. Use the blank radar chart (Workshop 1-1) and record your team's perception of your organization. You may need to exercise caution while evaluating your team's scores. High scores might reflect a lack of objectivity or honesty. Although low scores might seem discouraging, they may affirm the need for QFD. It is often interesting to have managers fill out another radar chart separately from the team members. You might discuss the differences early in a project to help define the boundaries of freedom for the team. Making contracts with management can provide a useful means of gathering support. Work on improving the environment before starting if you believe this is suitable. Depending upon what you learned, it may be appropriate to delay the start of your QFD project.

In this manual, workshops will be located in the outside column of the right-hand pages. A frame surrounds the worksheet and its instructions. The corresponding concepts are usually shown on the left-hand pages. Refer to the previous information for examples and clarification as you complete these exercises.

This book's documentation of QFD provides clear guidelines for anyone working on a future model or a similar project.

There is no magic to QFD, just plenty of intelligent, thorough work. The mill operations team leader at Kimberly-Clark provides an insightful commentary on the tools this book presents:

"QFD is not an easy process. It takes leadership and determination on the part of many people to dedicate the time and energy needed. But that effort pales in comparison to the effort expended in a poorly planned project" (Scheurell, 1993).

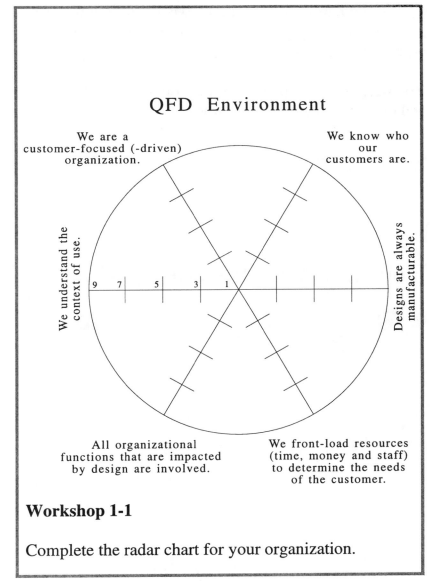

Workshop 1-1

Complete the radar chart for your organization.

Figure 1-9

Introduction

QFD is not a replacement for an existing design process; it is a tool that supports whatever design process an organization may be using.

You have now completed our general survey of QFD, and thus have begun your journey.

Your first steps towards QFD...

Chapter 2

Flow of Analysis in Step-by-Step QFD

Upon completing this chapter, you will be able to:

♦ Plan your QFD project.

♦ Integrate QFD into your design process.

♦ Create a different design process.

♦ Ask useful questions.

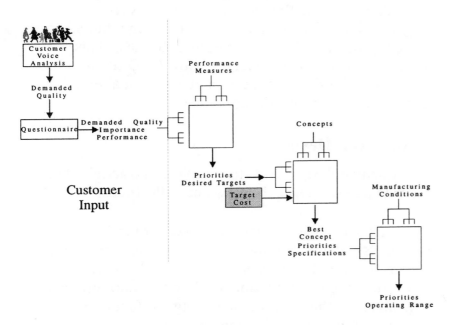

Step-by-Step QFD

for

Product Design

Customer
Input

15

Sequential Design

```
┌─────────────────────────────────┐
│     Identify the Market         │
└─────────────────────────────────┘
                │
                ▼
┌─────────────────────────────────┐
│    Select a Product Concept     │
└─────────────────────────────────┘
                │
                ▼
┌─────────────────────────────────┐
│        Design Product           │
└─────────────────────────────────┘
                │
                ▼
┌─────────────────────────────────┐
│     Confirm Product Design      │
└─────────────────────────────────┘
                │
                ▼
┌─────────────────────────────────┐
│      Design Manufacturing       │
└─────────────────────────────────┘
                │
                ▼
┌─────────────────────────────────┐
│   Confirm Design Manufacturing  │
└─────────────────────────────────┘
                │
                ▼
┌─────────────────────────────────┐
│      Manufacture Product        │
└─────────────────────────────────┘
```

Figure 2-1

Quality Function Deployment offers information that is not available through other processes. Just as modeling and market testing are necessary to answer certain questions, QFD activities invite organizations to formulate new questions that they have not considered before. The process may help an organization discover exciting information previously overlooked. These discoveries are a crucial part of connecting the voice of the customer to the design process. However, for QFD to be accepted and effective, it must become a part of an organization's existing design process. Every organization has its own approach to design, but all design processes share some basic activities. This chapter provides an overview of how QFD fits into your organization's design process and offers an alternative method for creating a design process.

Sequential Design

A sequential design tends to complete one activity before starting a new one, as the flow chart illustrates.

The first activity identifies the market or market segments. Identifying the market for an existing product means defining the market you intend to capture and understanding its requirements. For a technology-driven organization, this involves matching the functions and features of the new technology with a particular market. Such an organization could also work to create a market for the product. Sometimes a new technology "steals" market share from an older product. Currently, electronic mail and the Internet are taking over some of the market segments of FAX machines.

Once a business understands its market, it can begin selecting product concepts. Some criteria should be used to choose the best concept. Organizations joke about the design team tossing ideas over the wall separating it from manufacturing. Evidently, it is the manufacturing team's job to find a way to make a working version of the design. Designing for manufacturing is more effective because it requires communication between the design and manufacturing functions.

Product design includes simulations, bench testing, research and development. The manufacturer may build prototypes and try them out with small customer groups. Designed experiments may be necessary to optimize performance.

Eventually, the team finalizes and confirms a design as the one that will become the current model.

Depending upon the impact of the manufacturing process on previous and new designs, changes in the manufacturing methods may be necessary. Manufacturing may build pilot plants before up-scaling entire facilities. Sometimes larger batch sizes perform differently than smaller ones. At some point, the manufacturing process is confirmed.

If the design actually satisfies the customer and the manufactured product performs as intended, there will be no call backs or engineering changes after production startup.

Workshop 2-1

Brainstorm several possible products to be designed.

Select a product to be used in all the workshops.

Something simple is recommended, such as a pencil sharpener or mouse trap.

Buy several different types of the selected products in a local store.

Having several different brand names will greatly help discussion and stimulate ideas in Chapters 3, 4 and 5.

Figure 2-2

17

Concurrent Design

Concurrent design offers time compression and facilitates better communication during the design process. The staggered start for each of the activities requires communication across functions to ensure that each activity starts before preceding ones are finished. The cross-functional team, an integral requirement of QFD, becomes a necessity. Manufacturing actively participates in the early design activities. For technology-driven organizations, reducing the time between the birth of an idea and its implementation provides the greatest opportunity for time compression. This time period is often two or three times longer than the time allotted for introducing a new product.

Concurrent Design

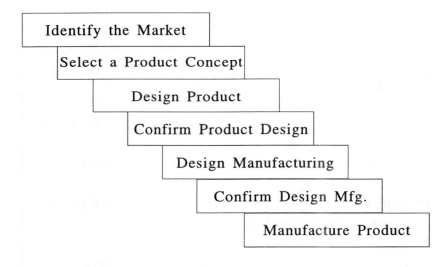

Figure 2-3

Product Design Process

Imagine that you work for a very unusual organization. Your superior has commanded you to do whatever is necessary to create a world-class design that outperforms the competition. Management wants to capture a large share of the market. Long-run financial returns are important. Money, time and resources are unlimited. Before you are given responsibility and authority for allocating resources and defining activities, you must make a presentation of your design process. Your supervisor expects this presentation to include a macro flow chart of about 20 steps.

You have already seen two macro flow charts in this chapter. Would there be any differences between the process you currently use and the one that would work best in this scenario? There would probably be several differences. First, there would be no time line with milestones for progress reports. Second, the most talented employees would have the time to become team members. Since they could avoid being distracted by other projects, team members would have no time conflicts.

There are, however, real time constraints. Time to market is often critical for market share. Some organizations predict lost sales as a function of time. Introducing a product today will yield the largest sales volume. Each day that passes after a competitor has introduced a product reduces the market share.

Workshop 2-2

Define a set of symbols to use in the flow chart. Your traditional sets are fine.

Identify about 20 process steps you would use for your team's design process.

Make a flow chart by linking these process steps.

Discuss and record how traditional constraints modify the implementation of this flow chart.

Figure 2-4

The Major Steps

This section covers five major steps in QFD. The first step (Figure 2-5) includes activities that focus on understanding the customer. The data produced are refined, and then a subset of the information becomes the input to the second step. The prioritization of the segments is one of the outputs from Step One. (Chapter 3 explores this process in detail.)

The analysis of the customer starts with identifying the customer segments, their characteristics and establishing the criteria used to prioritize the segments.

Step Two (Figure 2-6) involves gathering the voice of the customer and understanding the context in which the customers make statements. Contextual information clarifies the customer's verbatim information. The purpose of this activity is to establish a clear understanding of all of the customer's needs, specifically the subjective performance requirements. In this manual, these subjective performance requirements are called **demanded qualities**. Each QFD analysis builds in some way upon these demanded qualities. Since the design is driven by customer information, the practitioner is more likely to design a product that meets or exceeds the customer's expectations.

The demanded qualities become the foundation of a questionnaire for gathering more information about their importance and the current level of customer satisfaction with these items.

Figure 2-5

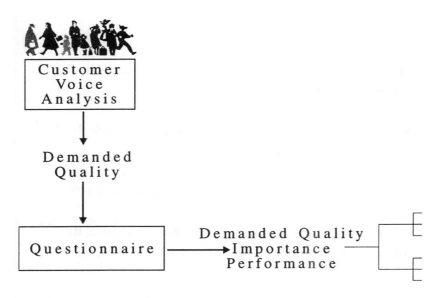

Figure 2-6

Step Three (Figure 2-7) translates the customer's statements and evaluations into the design team's performance measures language and priorities. The demanded qualities are the input to the matrix. If necessary, the design team uses demanded qualities to identify new concepts.

The design team sets priorities for demanded qualities by combining organizational priorities with the customer's priorities. The team also transforms the customer's subjective demanded qualities into technical performance measures, the output of the matrix . The team uses these **performance measures** to prioritize projects and establish desired target values for the design. These targets form the wish list that drives the design and development effort. Compromises may be required for the unattainable.

Step Four (Figure 2-8) employs Stewart Pugh's system for generating new concepts. Target costs are integrated in the generation process. The output of the previous matrix analysis becomes the input to this matrix analysis. The selected new concept and associated specifications are linked to the manufacturing process and the manufacturing database. The normal brainstorming process used to generate designs is significantly improved by using a method of systematic innovation called the Theory of Inventive Problem Solving (TRIZ or TIPS).

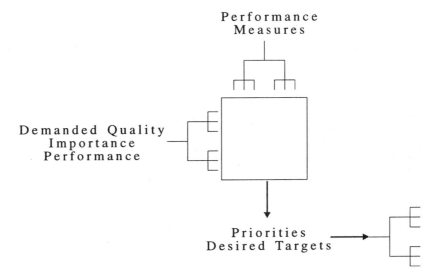

Figure 2-7

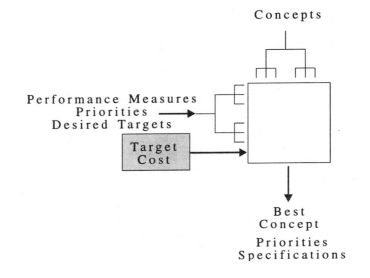

Figure 2-8

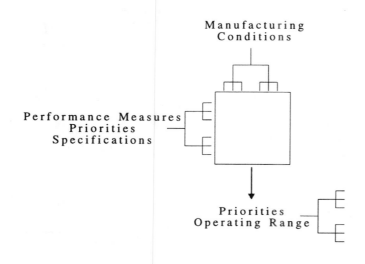

Figure 2-9

In Step Five (Figure 2-9), the final analysis uses a matrix to link the product specifications to the manufacturing conditions. Identifying the knowledge base for the relationship between operating conditions and product performance is part of the manufacturing database. The output of this analysis can be control systems or procedures.

QFD does not replace an organization's existing design process. Accordingly, you can integrate QFD into a sequential, concurrent or your unique design process. For example, consider a subset of the sequential design process flow chart and the five major QFD activities. Note that you can introduce a QFD activity at various points in the design process. The approach is flexible enough that your design team can decide when to start and stop the QFD process.

Any single QFD activity will aid your understanding of the input you are using. You could merely apply QFD to the manufacturing process. You may produce a product that the customer does not want, but you will inevitably improve the manufacturing process. In general, using a middle phase of QFD will enhance your understanding, but you will gain more if you combine the activity with one or some of the earlier activities. As a rule of thumb, using any of the earlier activities is more effective than choosing one of the later phases.

Design Process

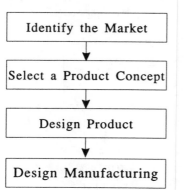

QFD Contribution

Prioritize Customer Segments (Chp 3), Understand Customer Needs and Context (Chp 4), Translate into Engineering Language (Chp 5) Select the Best Concept Generate New Concepts Target Cost (Chp 7)

Prioritize Development Projects Establish Targets (Chp 5)

Establish Relationships between Manufacturing Conditions and Product Performance (Chp 9)

Figure 2-10

Developing a Design Process

It is necessary and important for an organization to improve, standardize and maintain the product development process. Accordingly, an effective organization should do the following:

- Improve the design process continuously.
- Review missing design processes.
- Know resource allocations for successful products.
- Review and develop cross-functional activities.
- Manage the product development process.

Structuring a flow chart for all development and implementation activities is a recommended procedure. Structuring the design activities is the next step in the planning process.

QFD practitioners use the PDCA (Plan, Do, Check, Act) cycle to create a design flow. This cycle reveals the relationship between a design flow and organizational functions. This approach introduces a different way of looking at the design process that often generates new, useful insights.

The classic definition of PDCA is:

Plan - create a model to be tested

Do - try the model

Check - compare actual results and predicted results

Act - modify or solidify the theory

PDCA is a critical process. Using the PDCA cycle in a

design process requires adapting the classic definitions of these terms.

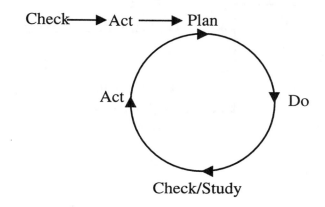

Figure 2-11

The P (Plan) integrates and translates all relevant data from the voice of the customer into design requirements. The PDCA planning step also defines several additional activities at a finer level of detail.

The D (Do) includes integrating technologies and concepts and confirming that the design satisfies the design requirements.

The C (Check) verifies the actual product performance.

Any differences between the team's expectations and the product's actual performance are used in the A (Act) step to decide whether the design cycle is completed or whether modifications must be made to correct deviations in C.

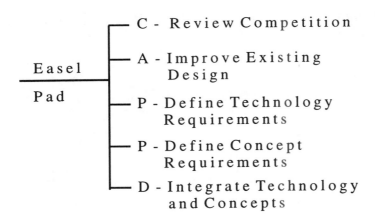

C - Review Competition

A - Improve Existing Design

Easel Pad

P - Define Technology Requirements

P - Define Concept Requirements

D - Integrate Technology and Concepts

Figure 2-12

Critical Process Responsibility ⬤ Primary ◯ Supporting	President	Marketing	Engineering	Research & Design	Quality	Manufacturing
C - Review Competition	◯	⬤	◯		◯	
A - Improve Existing Design		◯	⬤	◯	◯	◯
P - Define Technology Requirements			⬤			◯
P - Define Concept Requirements		◯	⬤	◯	◯	◯
D - Integrate Technology and Concepts		◯	⬤		◯	⬤

Figure 2-13

Developing the Critical Process

The PDCA cycle is a proactive process. A reactive cycle identifies problems after they have occurred. A better strategy would be to look and think before starting the PDCA cycle. Looking and thinking correspond to Checking and Acting. Proceeding with C and then A forms a reactive cycle. This CAPD cycle starts by checking the current performance of existing products.

The CAPD cycle can be used to identify the first-level tasks needed to design an easel pad. The critical process for planning the design of the easel pad is contained in the branches of the tree in Figure 2-12. Within the CAPD cycle, there are several tasks that must be completed. Each element of CAPD must have at least one task. For instance, the Plan portion should include defining technology and concept requirements. This process may identify additional tasks besides those that have been part of the traditional design process. Though they may have been performed informally, these tasks were missing from the formal process.

Responsibility for the Critical Process

Responsibilities must be assigned to the appropriate organizational functions. A responsibility matrix is useful for summarizing these responsibilities. The solid circle signifies primary responsibility; the open circle, secondary responsibility. A space indicates no responsibility. The functional group which has the primary responsibility for a task determines when to move on to the next step in the critical process. The functional group which has secondary responsibility provides support to the group shouldering the primary responsibility. Job position or task description identifies who is responsible for which tasks. According to the developed critical process, an organization may need to acquire new responsibilities and to divide these among the functional units. The pattern of circles in Figure 2-13 also indicates that the engineering department may have too many responsibilities.

Identifying Critical Tasks

The CAPD cycle is used to identify the critical tasks for each of the major tasks in the critical process. These first-level tasks in the CAPD cycle require at least one second-level task. There are five tasks for the first major task in the critical process (Figure 2-14):

- C - Identify target market
- C - Identify customer needs and expectations
- A - Select performance measures
- P - Design technical benchmarking
- D - Evaluate competition

Depending upon the steps in the critical process, the critical tasks can be different. For more complex design processes, it may be necessary to use the CAPD cycle again to identify critical subtasks.

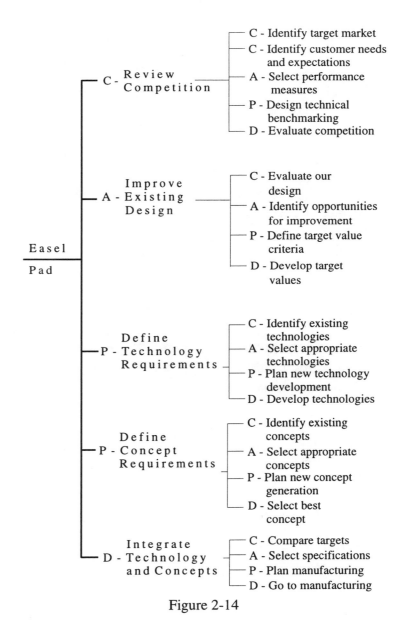

Figure 2-14

Task Responsibility ● Primary ○ Supporting	President	Marketing	Engineering	Research & Design	Quality	Manufacturing
C - Identify target market	●	●	○			
C - Identify customer needs		●				
A - Select perform. measures		○	●	○	○	
P - Design tech.bench market		●	●		●	
D - Evaluate competition		○	●			
C - Evaluate our design			●	●		
A - Identify impr. oppor.		●	○	○		○
P - Define target value criter.		●				○
D - Develop target values		●	●	○	○	○
C - Identify exist. tech.			●	●		
A - Select appr. tech.			○	●		○
P - Plan new tech. devel.			○	●		
D - Develop technologies			●			
C - Identify exist. concepts		○	●	●		
A - Select appr. concepts		○	●	●		○
P - Plan new concept gen.			●			
D - Select best concept		○	●	○	○	○
C - Compare targets		●	●			
A - Select specifications			●			●
P - Plan manufacturing			○			●
D - Go to manufacturing						●

Figure 2-15

Responsibility for the Critical Process

Using the responsibility matrix for the critical process (Figure 2-13), an organization identifies the functional responsibility for each task (Figure 2-15). The more detailed responsibility matrix uses the same symbols.

The Product Design Process Chart

The Product Design Process Chart is shown on the next page. Rows contain the critical processes and critical tasks. Columns represent the organization's functional units. Rectangles in the flow chart identify activities and location signifies participation in the activity. Arrows indicate the flow of documents or decisions. The column in which an output arrow is placed indicates primary responsibility.

The Product Design Process Chart clearly defines team membership. During product development, the design team needs many documents, such as market analyses, customer requirements and specifications. These documents are normally regarded as reports within the design process. For the sake of clarity, these documents have been excerpted and only the QFD reports are placed in the right-hand column of Figure 2-16. For this example, the column contains four of the five QFD topics covered in this manual. The manual presents the customer voice analysis in two steps—gathering the customer's demanded qualities and then prioritizing these needs.

Product Design Process Chart

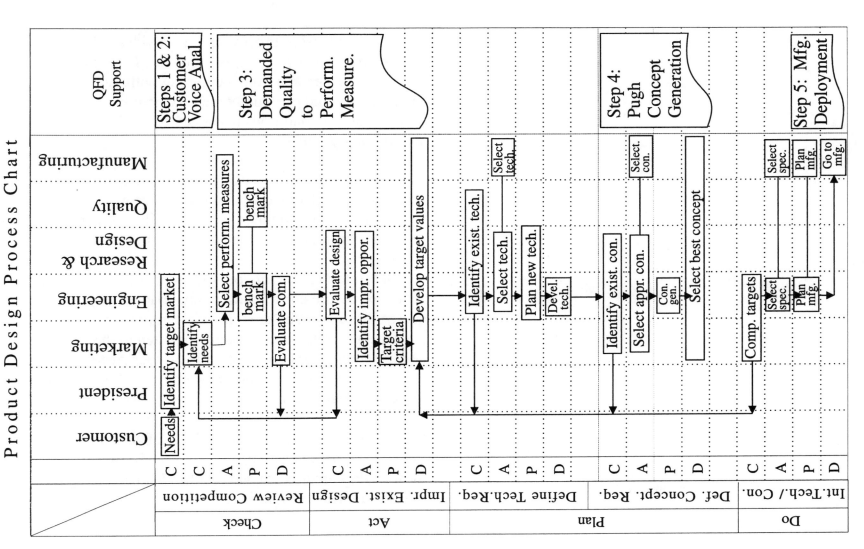

Figure 2-16

Figure 2-17 presents the sequence of the QFD activities included in this chapter.

The Product Design Process Chart provides a detailed road map for managing a QFD project. The journey unfolds, one step at a time, beginning with the voice of the customer and eventually identifying the manufacturing conditions necessary to produce a world-class product.

It is possible that simply using the PDCA cycle to construct a Product Design Process Chart will provide a significant breakthrough in thinking. This new road map to the design process may be all the team needs.

It is helpful to reflect on your different customer segments before considering future changes in your design process. Who are your most important customers? How much should you listen to the demanded qualities voiced by one of your customer segments? The next chapter presents a ranking system for answering these questions.

Step-by-Step QFD includes a variety of activities that can be intermediate end points. The team can stop the formal QFD process whenever it has learned enough to effect a significant benefit in the design process.

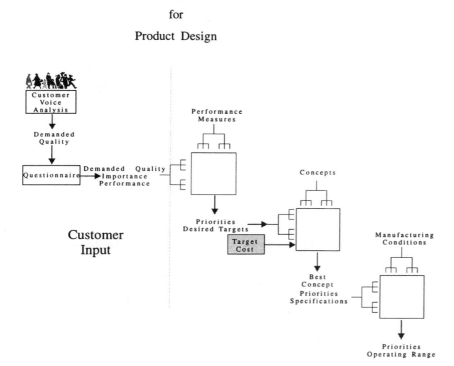

Step-by-Step QFD

for

Product Design

Customer
Input

Figure 2-17

Managers can show their support for these activities by asking some of the following questions:

1. Which functions are represented on the team?

2. What customer segments have been considered?

3. What criteria have been used to rank customer segments?

4. What breakthrough in thinking has resulted from experiencing the customer's environment?

5. Are there any previously unknown demanded qualities?

6. Have there been any surprises in customer importance and evaluation of the current design?

7. Are there any performance measures that are no longer needed?

8. Are there any new performance measures?

It is time to begin the QFD journey.

Chapter 3

Priority of Customer Segments

Upon completing this chapter, you will be able to:

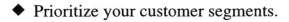

◆ Prioritize your customer segments.

◆ Use the Analytic Hierarchy Process (AHP).

Figure 3-1

Customer Segments

Who should influence your product design?

Stakeholders include anyone who can influence the decision to use or buy the product and anyone who is impacted by the use of the product. The list of stakeholders can be substantial, and your design may require listing all of their needs. This process often helps identify important overlooked customers.

Who is the customer for dog food? This simple inquiry may seem silly, but a true story shows what can happen if we fail to ask this question. Several years ago, a dog food manufacturer designed the perfect product. The design team included animal nutritionists, veterinarians, marketing and manufacturing personnel, dog show judges and owners of working dogs. The sales campaign was supported by recognized experts and they promoted the product as a balanced diet. Sales during the first month exceeded the most optimistic forecasts. Production capacity was increased, but sales were near zero during the second month. The design team had forgotten an important stake (steak) holder: the dogs would not eat this wonder food.

Who is the customer for the soaps and shampoos that are found in the bathrooms of hotels and motels? In 1986, Procter & Gamble increased its market share by 400% for bathroom toiletries used in hotels and motels. The company had historically believed that the hotel guest was its primary customer. However, after listening to hotel managers, Procter & Gamble realized that these stakeholders wanted the products to communicate their particular hotel's image. Procter & Gamble redesigned the packaging, concentrating on image. Without changing the product's content, it increased market share by 400%. Prior to this decision, the toiletries looked like samples from the local discount house. After recognizing the image need of the hotel, Procter & Gamble incorporated floral patterns and other decoration into the package design. While the new product line does not clearly display Procter & Gamble's logo, its packaging now fits its function, communicating the hotel's image.

Who are the stakeholders for the product identified in Workshop 2-1?

Who are the stakeholders for your organization's products?

Which customer segments are particularly important to your design?

Workshop 3-1

Who are the stakeholders for the product identified in Workshop 2-1?

Use the brainstorming process to identify all the possible customers and users of your product.

Use the brainstorming process to identify all the individuals and organizations that can influence the decision to buy your product.

While brainstorming, keep asking, "What is a stakeholder?"

Figure 3-2

Measurement

A total quality management (TQM) organization makes decisions based upon facts. This chapter, and the remaining chapters, will utilize all the information contained in any data. Using arithmetic calculations to prioritize your customer segments and the importance of their needs is an integral part of QFD.

Nominal
mode

Ordinal
mode, median

Interval
mode, median, mean

Ratio
multiply/divide

Figure 3-3

Practitioners must be cognizant of the levels of their measurement systems. There are four levels of measurement, each one containing more information than its predecessor.

The nominal scale is only an identifier. This type of data is mathematically sound for counting and calculating the mode. The mode is the number that occurs most frequently.

Numbers in the ordinal scale depict positioning in a hierarchy. Ordering the numbers has meaning. However, it is not appropriate to say 4 is twice as large as 2 or that the difference from 2 to 4 is the same as the distance from 10 to 12. Use this system to find both the mode and median (the middle value).

In the interval scale, the difference between the numbers has meaning. With this measure, the difference between 2 and 4 is the same as the difference between 10 and 12. All three of the measures of central tendency (mean, median and mode) are mathematically available. The mean is also called the average. Ratios are not meaningful for the interval scale.

The ratio scale includes an absolute zero. With the absolute zero, finding ratios and multiplying numbers becomes meaningful. A few academics have questioned some calculations typically used in QFD because ratios and products are found without having ratio data.

Since the numbers associated with time, distance, volume, work, energy and many other factors are all ratio data, it is tempting to treat any number as ratio data, even when it is not appropriate. To avoid this error, this manual offers alternative ranking systems, including the Analytical Hierarchy Process (AHP). This process creates ratio data for ranking.

The design process for most organizations is so flawed that current violations of the requirements for ratio data still yield significant improvements in decision making. However, as competitors become more experienced in the QFD process, the ability of an organization to use ratio numbers in its analyses will distinguish the world-class providers from the others.

Ranking

All five of the following ranking systems are used within QFD. As practitioners and their competition become more experienced, their methods evolve toward full pairwise comparisons.

The 1 to 5 scale is frequently used in evaluation questionnaires. A customer may be asked to rank a product's performance along several dimensions, such as *ease of use* and *appearance*. The customer assigns a 1 if the performance is unacceptable and a 5 if it is world class. These numbers are most likely only ordinal data, but they are mistakenly treated as ratio data by most practitioners.

A 1 to 9 scale is more repeatable, based upon research by Thomas Saaty. The nine-point scale is better than an eight- or ten-point scale. It is more of a challenge for the customer than the five-point scale, but allows for more gradations.

A ranking process works better than the simple 1 to 5 and 1 to 9 systems. First, all of the competitors for one measure of performance are arranged in ascending order by customer satisfaction. Then a number is assigned to indicate the relative distance between each competitor. There is more information contained in these numbers, but they are not quite interval data.

Ranking Systems

1. 1 -- 5

2. 1 -- 9

3. Ranking

4. Distributing 100 Points

5. Full Pairwise Comparison

Figure 3-4

Relation \ Age	Young	Middle	Old
Stranger			
Friends	Students		Senior Citizens
Lovers	Lovers		XXX
Family	Young Family		

Figure 3-5

	Age	Education	Job Function	Income	Industry
1	1-5	None	Learning	none	Manufacturing
2	6-14	High School	Teaching	100	Education
3	15-20	College	Changing Behavior	2000	Defense
4	20-40	Private	Purchasing	50,000	Research
5	40-99	Trade School	Designing	100,000	Consultation

Figure 3-6

Consumer surveys use a fairly common version of the ranking approach. First, the customer distributes 100 points among several groups of characteristics. Then, for each characteristic, the assigned points are further distributed among the members of the group. For example, if the project is designing a new computer monitor, some of the groups might be: *easy to adjust*, *good resolution* and *compatible with all software*. The allocation of importance points could be 20, 30 and 50, respectively. The group *good resolution* could contain: *sharp colors*, *many density options*, *full range of colors* and *works in any lighting*. The distribution of the 30 importance points could be 5, 5, 2 and 18, respectively. These numbers offer a fairly close approximation of ratio data; however, the process is challenging for the customers.

A full pairwise comparison using a nine-point scale will yield ratio data. The approach presented is called the Analytic Hierarchy Process. AHP is backed by 30 years of research. Ranking only two items is the easiest system for the human brain. The customer may tire of comparing all pairs of combinations for the list of items to be ranked. Nevertheless, the precision of the AHP process is well worth the effort.

AHP has been used in a variety of applications, from selecting the site for the new Sudan Airport to establishing the criteria for selecting a college and then ranking the various college options.

This chapter focuses on using AHP to rank different customer segments. The first step is determining the different characteristics that can identify the particular customer segment to which an individual belongs.

A restaurant is used as an example to introduce the concept. If you were going into the restaurant business, four possible customer segments are Senior Citizens, College Students, Lovers and Young Families. Some characteristics that sort individuals into these categories are Age, Occupation, Marital Status and Relation to Others While Eating. The type of restaurant planned would influence the choice of characteristics you should consider. For instance, the characteristics for a student segment may not merit lengthy deliberation if you plan to open an expensive restaurant. At the same time, brainstorming often works best without such constraints. To generate a more comprehensive list of possible characteristics, you may choose not to ignore any customer segments.

Brainstorming to consider different combinations of all of the characteristics helps to identify overlooked, hungry customer segments. A matrix is used in Figure 3-5 to look at two of the characteristics, Relation and Age. The currently identified segments are entered in the cells. Conventional market segments can be assigned to these options and previously overlooked customer segments are identified in the blank combinations. Looking at all the combinations of Age and Relation in Figure 3-5 identifies nine combinations not considered (blank). One example is senior citizen lovers, as indicated by the XXX. Is there a market for such a group? It could work in a retirement community.

For the easel pad, customer segments Teachers, Students, Purchasing Agents and Consultants have been selected. It is always important to make sure nothing or nobody is forgotten. In this case the morphological table is useful. If an unknown person is to be assigned to one group,

what questions might we ask. These will be called characteristics for identifying customer segments. Age, years of school, job function, income, industry are used as an example.

These are then placed in a table as column headings in Figure 3-6. Possibilities for each column are entered. The following is just a sample. For your application everything can be completely different.

For example

Some Teachers are in the combination 43242

Some Students are in the combination 32112

Some Purchasing Agents are in the combination 53451

Some Consultants are in the combination 54333

The total number of segments is 5x5x5x5x5x5=3125. There must be some interesting segment contained in that set that you had not considered and offers a breakthrough in marketing. Obviously for more columns with longer lists the complete enumeration is not feasible. Patterns become clear as you work with the list for awhile. Children ar home could be another market segment for the easel pad.

Characteristics

1	5
2	6
3	7
4	8

Workshop 3-2

Make a morphological table.

Brainstorm a list of stakeholder characteristics for the product you have chosen for your project.

For the purposes of this workshop, select four characteristic and identify several options for each.

Identify several combinations of characteristics and the related customer segements.

Figure 3-7

Examining characteristics allows organizations to uncover new market segments. Once customer segments are identified, it is necessary to determine the relative importance of each to decide who to talk to and how much weight to assign to their opinions. Most organizations have their own criteria for evaluating the relative importance of different customers. These criteria should be used formally in ranking different segments. The criteria Market Size, Cost to Support, Easy to Satisfy and Publicity (the ability to improve sales by word of mouth) will be used to show four different types of criteria. If Market Size is 100 times more important than the other criteria, then the customer segment with the largest market size would be the most important. Ranking the importance of market segments is challenging when more than one criterion is significant in decision making. A weighted score for the customer segments is the solution. The AHP is used to calculate ratio data for ranking the criteria.

The Analytic Hierarchy Process

While he was a member of the Wharton School of the University of Pennsylvania, mathematician Thomas L. Saaty developed the Analytic Hierarchy Process (AHP) to weight concepts, products, options and other important items. This multicriteria decision-making process enables you to work with both tangible and intangible factors. The process employs a 1 to 9 weighting scheme for paired comparisons.

Two-way comparisons are the easiest for the human brain. A consistency check can be made. If A is twice as desirable as B and B is twice as desirable as C, then to be consistent, A must be four times as desirable as C. The paired comparisons in AHP are appropriate for a maximum of 7 ± 2 items.

The focus for weighting in AHP can be on the basis of importance, preference or likelihood; the weights can be verbal or numerical:

Importance is used for comparing one criterion with another.

Preference is used for comparing alternatives.

Likelihood is used for comparing the probability of outcomes and is appropriate for criteria or alternatives.

Process will be introduced by ranking the criteria. A matrix for all the paired comparisons for criteria is shown in Figure 3-9. Each row is compared to each column using the nine-point scale of importance in Figure 3-8. The even numbers have importance between the defined odd numbers. This nine-point scale is worded for when the row is more important than the column.

1. Equal importance: the row and column have the same impact upon the higher order need.

2. Between 1 and 3.

3. Moderate importance: experience and judgment slightly favor the row over the column.

4. Between 3 and 4.

5. Strong importance: experience and judgment strongly favor the row over the column.

6. Between 5 and 7.

7. Very strong importance: the row is strongly favored and its dominance is demonstrated in practice.

8. Between 7 and 9.

9. Extreme importance: the evidence favoring the row is of the highest possible order of affirmation.

Figure 3-8

Priority of Customer Segments

	Market Size	Cost to Support	Easy to Satisfy	Publicity
Market Size	1.00	5.00	9.00	3.00
Cost to Support	/	1.00	2.00	/
Easy to Satisfy	/	/	1.00	/
Publicity	/	2.00	3.00	1.00
Total				

Figure 3-9

Start with the question "Is this row more important than or equal to the column?"

If the answer is YES, use the nine-point scale. If the answer is NO, place a divide symbol (/) in the cell.

	Market Size	Cost to Support	Easy to Satisfy	Publicity
Market Size	1.00	5.00	9.00	3.00
Cost to Support	0.20	1.00	2.00	0.50
Easy to Satisfy	0.11	0.50	1.00	0.33
Publicity	0.33	2.00	3.00	1.00
Total	1.64	8.50	15.00	4.83

Figure 3-10

In row 1 (Figure 3-9), Market Size is just as important as Market Size for ranking the customer segments; therefore, 1.00 is entered in the cell where the row and column meet. A value of 1.00 is entered in each cell along the diagonal from the upper left to the lower right, as these cells compare one criterion with itself.

In row 1, Market Size is more important than Cost to Support for ranking the customer segments. A value of 5.00 is entered because the team agreed that Market Size is significantly more important than Cost to Support as scored in Figure 3-8. If the row is not more important than the column, a divide symbol (/) is entered. The process is continued for all cells.

Since Market Size is considered five times more important than Cost to Support, the entry that shows how important Cost to Support is compared to Market Size must be 1/5 or 0.20 (Figure 3-9). This process is continued for all the entries with (/).

Figure 3-10 shows these calculations and column totals. Note that the cells below the diagonal of 1.00's, going downward from left to right, are the reciprocals of cells above the diagonal.

Any ranking system, including AHP, only works if the items being compared are of the same order of magnitude. Comparing the light of the sun with a spotlight and a fluorescent ocean alga is not a fair comparison. When the comparison is fair, the numbers in the nine-point scale can be used to evaluate the relative significance of one item to another.

When working with a customer, use the verbal form of the nine-point scale (Figure 3-8) and not the numbers for ranking.

The columns are normalized in Figure 3-11 (finding the fraction of the total for each column). The average of the rows gives the ranking of the customers for the one criterion. These average results are entered in the last column.

	Market Size	Cost to Support	Easy to Satisfy	Publicity	Total	Average
Market Size	0.608	0.588	0.600	0.621	2.417	0.604
Cost to Support	0.122	0.118	0.133	0.103	0.476	0.119
Easy to Satisfy	0.068	0.059	0.067	0.069	0.262	0.066
Publicity	0.203	0.235	0.200	0.207	0.845	0.211
Total	1.000	1.000	1.000	1.000	4.000	1.000

Figure 3-11

You can combine all these calculations in a single table (Figure 3-12). Double entries in the cells give the raw rankings in the upper left and the normalized scores in the lower right. The rankings of importance for criteria Market Size, Cost to Support, Easy to Satisfy and Publicity are 0.604, 0.119, 0.066 and 0.211, respectively. Market Size is nearly three times more important than Publicity.

Workshop 3-3 uses the combined table format.

	Market Size	Cost to Support	Easy to Satisfy	Publicity	Total	Average
Market Size	1.00 / 0.608	5.00 / 0.588	9.00 / 0.600	3.00 / 0.621	2.417	0.604
Cost to Support	0.20 / 0.122	1.00 / 0.118	2.00 / 0.133	0.50 / 0.103	0.476	0.119
Easy to Satisfy	0.11 / 0.068	0.50 / 0.059	1.00 / 0.067	0.33 / 0.069	0.262	0.066
Publicity	0.33 / 0.203	2.00 / 0.235	3.00 / 0.200	1.00 / 0.207	0.845	0.211
Total	1.64 / 1.00	8.50 / 1.00	15.00 / 1.00	4.83 / 1.00	4.000	1.000

Figure 3-12

Criteria

1	4	7	10
2	5	8	11
3	6	9	12

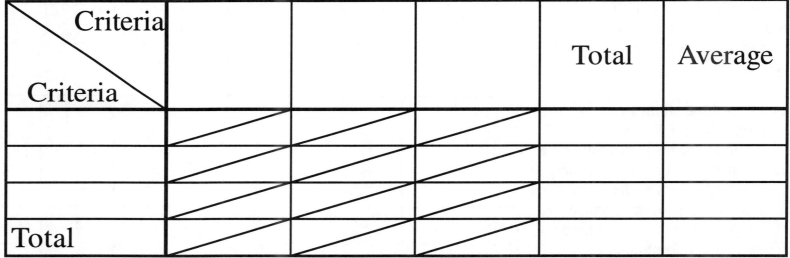

Criteria \ Criteria				Total	Average
Total					

Workshop 3-3

Brainstorm possible criteria for ranking your selected customers.

Select three criteria that you will use to rank your customers.

Use AHP to rank your three selected criteria.

Figure 3-13

42

Information can be gathered on Market Size, Cost to Support and Easy to Satisfy. The data for criteria may be classified as either measurable or subjective. The measurable data can reflect larger is better, smaller is better or an expert opinion. It is generally unproductive to gather data for the unimportant criteria.

Market Size represents how many units each segment is expected to buy; the larger the number, the more desirable the customer. The Consultants, with 90,000 units, are ten times more desirable than the Teachers, with 9,000.

The smaller the Cost to Support, the more desirable the customer. In this case, the Teachers are seven times more desirable as customers than the Consultants.

For Easy to Satisfy, the ranking of 10 for the Students means they are 2.5 times easier to satisfy than the Teachers, with a ranking of 4. This type of data comes from expert judgments, such as restaurant ratings of a ★★★★ restaurant compared to a ★★1/2 restaurant. At first glance, such judgments may seem subjective, but these data are standardized insomuch as they reflect the professional opinions of experts. Similarly, the opinions of professional wine tasters and perfume testers are treated as legitimate measurements.

	Teacher	Student	Consultant
Market Size	9000	1000	90000
Cost to Support	$1/	$2/	$7/
Easy to Satisfy	4	10	1
Publicity	?	?	?

Figure 3-14

The ability of the different customer segments to promote the sales of the easel pad involves a judgment ranking. The AHP process will be used to rank each customer segment's ability to promote the easel pad's sales.

Each of these criteria requires a different transformation before it can be combined with the criteria weights into a weighted priority of the customer segments.

Priority of Customer Segments

	Teacher	Student	Consultant	
Market Size	9000	1000	90000	market units
	0.09	0.01	0.90	fraction of total

Figure 3-15

The number of units sold is a measure of Market Size. Calculating the percentage of the total market or the fraction of the total is appropriate because counted numbers are ratio data. The total is 100,000 units. Thus, 9% or 0.09 of the total anticipated sales will be from Teachers. The Consultants are 10 times more desirable than the Teachers, the same as before.

	Teacher	Student	Consultant	
Cost to Support	$1/	$2/	$7/	$ per person
	1.000	0.500	0.143	reciprocal
	0.61	0.30	0.09	fraction of total

Figure 3-16

Because a smaller number is better for Cost to Support, a two-step process is necessary. The reciprocal is found first. The second step calculates the total for the reciprocals and is used to rank each customer as a fraction of the total.

	Teacher	Student	Consultant	
Easy to Satisfy	4	10	1	ranking
	0.27	0.67	0.07	fraction of total

Figure 3-17

Easy to Satisfy is a rating system and is changed to a fraction of the total satisfaction.

AHP will be used to create ratio data for the ranking of the customer's ability to promote the sales of the repositionable easel pad (Publicity). The Publicity data are not known; thus, the data for this criterion are subjective. Our best judgment will be used and supported by making ratio data for these judgments. Your organization can use its best judgment by making paired comparisons between different customer segments and the criterion Publicity. Use the 1 to 9 scale in Figure 3-8 to ask the question:

How much more important is customer segment X than customer segment Y for helping Publicity?

The average column shows the rankings for the customer segments.

Customer / Customer	Teacher	Student	Consultant	Total	Average
Teacher	1.00 / 0.18	2.00 / 0.16	0.25 / 0.18	0.532	0.177
Student	0.50 / 0.09	1.00 / 0.08	0.11 / 0.08	0.256	0.085
Consultant	4.00 / 0.72	9.00 / 0.75	1.00 / 0.73	2.212	0.737
Total	5.50 / 1.00	12.00 / 1.00	1.36 / 1.00	3.000	1.000

Figure 3-18

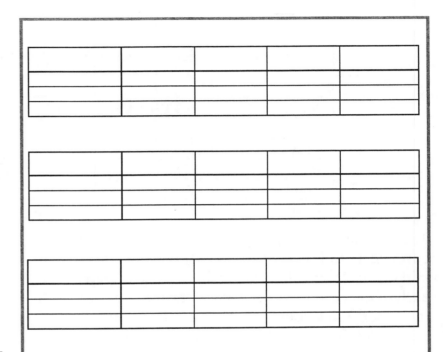

Workshop 3-4

Determine which of the four criteria types each of your criteria belong to (larger-is-better, smaller-is-better, expert opinion or AHP, as shown in Figures 3-16 to 19).

Rank your three customers for each of the three criteria you selected (Workshop 3-3), as was done for the easel pad example.

Use the appropriate method for calculating the fraction of the total score.

Figure 3-19

Priority of Customer Segments

Criterion	Priority	Teacher	Student	Consultant		Teacher	Student	Consultant
Market Size	0.604	0.090	0.010	0.900		0.054	0.006	0.544
Cost to Support	0.119	0.609	0.304	0.087		0.072	0.036	0.010
Easy to Satisfy	0.066	0.267	0.667	0.067		0.017	0.044	0.004
Publicity	0.211	0.177	0.085	0.737		0.037	0.018	0.156
Total	1.000				Importance	0.182	0.104	0.714

Figure 3-20

The weighted priority can now be calculated for the customers. In Figure 3-20, the left two columns have the criteria and their calculated weights. The importance of each of the customers for each of the criteria is recorded in the next three columns. The customer importance for each criterion is from Figures 3-15, 3-16, 3-17 and 3-18, respectively.

The weighted importance for each segment for each criterion is the product of the importance of the criterion and the importance of each customer segment for that criterion. The product of the importance of Market Size (0.604) times the ranking of the Teacher for this criteron (0.090) is the highlighted 0.054 in the Teacher column to the right. The other calculations are recorded in the three columns to the far right.

The column totals are the weighted importance for each of the customers. The results for the Teacher, Student and Consultant are 0.182, 0.104 and 0.714 respectively. Because the Market Size criterion is so important and the Consultant is the most desirable customer for this criterion, the Consultant emerges as the most important customer overall. (Chapter 5 presents three ways of satisfying multiple customers.)

Rank your customer segments in the next workshop using the double entry table.

Criterion	Priority	Teacher		Student		Consultant	
Market Size	0.604	0.090	0.054	0.010	0.006	0.900	.0.544
Cost to Support	0.119	0.609	0.072	0.304	0.036	0.087	0.010
Easy to Satisfy	0.066	0.267	0.017	0.667	0.044	0.067	0.004
Publicity	0.211	0.177	0.037	0.085	0.018	0.737	0.156
Importance			0.182		0.104		0.714

Criterion	Priority			
Importance				

Workshop 3-5

Calculate the weighted importance for your three customer segments. The information in columns 1 and 2 comes from Workshop 3-3. The information in columns 3, 4 and 5 comes from Workshop 3-4.

Figure 3-21

If you are unfamiliar with your criteria, comparing the customer segments for each criterion first will clarify how your team is choosing to standardize your interpretation of the criteria. The criteria would then be ranked. The process for determining the weighted priorities of the customer segments would be the same.

This chapter presented a brief case and some artificial criteria for illustrative purposes. The process will work for larger cases. As mentioned earlier in this chapter, a maximum of 7 ± 2 criteria or options produces the most repeatable results. If there are more than nine criteria, they must be grouped to form a smaller set. Chapter 4 offers a method for creating these smaller groupings.

You can now prioritize your customers and projects. You can even use this process for purchasing your next automobile or selecting a spouse!

For a more detailed presentation of AHP, consult Saaty's book *Decision Making for Leaders*. The text explains tests for consistency and other applications of AHP.

Some organizations find that the prioritization of their customers provides enough new insights. They do not continue with the QFD process. Instead they return to their traditional design process without the additional input of QFD. For those continuing up the mountain, the next chapter takes a detailed look at the customer's needs and desires.

It is time to

and review your understanding.

You have taken another step toward better design!

現場

```
┌─────────────┐
│  Customer   │
│   Voice     │
│  Analysis   │
└─────────────┘
       │
       ▼
  Demanded
  Quality
       │
       ▼
┌─────────────┐      Demanded   Quality
│Questionnaire│─────▶Importance
└─────────────┘      Performance
```

Chapter 4

Understanding Your Customer

Upon completing this chapter, you will be able to:

♦ Document the environment where your product is or will be used.

♦ Expand the list of customer-stated needs.

♦ Sort the customer-stated needs into demanded qualities, performance measures, functions and solutions.

♦ Determine the natural clusters of demanded qualities.

♦ Calculate the importance for the demanded qualities.

Voice of the Customer

The purpose of this chapter is to further your understanding of the customer's needs, dreams, wishes and expectations. The context of all of these will influence the design process.

Once you have identified the customer, it is time to walk in his/her moccasins. This North American Indian expression serves as a reminder of the importance of trying to understand other people's motivations and behavior. If you lived the lives of the customers, perhaps you would behave as they do and understand their needs. Many organizations make a concerted effort to walk in their customer's moccasins; some even require design team members to work in the customer's organization for

INFO ABOUT PERSON	VOICE OF CUSTOMER
53 years	I can move it
Consultant	works on my walls
1.9 m tall	
12345A	

Figure 4-1

six months. For example, a tractor engine manufacturer assigned engineers to live and work with farmers for one year.

Taking the time to know your customer's needs requires front-loading of money, time and personnel. This is why one dimension of the Radar Readiness Chart in Chapter 1 contains the "front-loading" of resources. The traditional resource allocation starts very modestly and grows almost exponentially. QFD requires heavy front-loading if the customer is to drive the design or at least set the focus of the design effort.

The Japanese expression "going to the *Gemba*" provides a vivid image of understanding your customers. The kanji characters for *Gemba* present an integration of the symbols for a king who is overlooking the land and watching swine herds working under the sun. The first Japanese character combines the king overlooking, while the second contains land and the swine herds laboring under the sun. This picture of scrutinizing the people you are serving provides a banner for this chapter's purpose. This banner encapsulates the importance of understanding the customer's environment and his/her behavior in that environment.

現場

Has your organization gone to the *Gemba*? The first step in your journey to the *Gemba* entails talking to the customers and recording what they say. Document whom you are interviewing or observing in the first column (Figure 4-1). Often this information is of a demographic nature, such as income, age or education. This data also includes an identification number used later to locate the original source during team discussions. Record the customer's verbatim response rather than the solution to the response. The exact words must be recorded, and it is crucial to include any information about the customer that would impact product design or how the customer would use the product. Is the customer right- or left-handed? What is his profession? In our case, the customer is identified as a 53-year-old consultant who is 1.9 meters tall. Some team members may see the verbatim response and say, "How could he want that? It does not make any sense." But it does make sense to the customer. This person wants to do things his way, not your way.

The consultant is the customer in this example (Figure 4-1). Only two statements have been recorded. In a real example, there are several statements. The consultant wants to be able to move the easel pad sheets and have them stick to the surfaces in his environment.

Some organizations find that the information gained through this approach is already so different from previous inquiries that they decide to stop the QFD process and resume their traditional design procedures. The difference is a new understanding of the customer, causing a major breakthrough in concepts appropriate.

It is time to and review your understanding.

INFO ABOUT PERSON	VOICE OF CUSTOMER
	A

Workshop 4-1

Use worksheet A from the back of this manual.

Select one customer for this workshop.

Record information about the customer that affects product design and how the customer might use the product. Consider factors such as education, profession, age or height. Have you ever watched a very tall or very short person remove a sheet of paper from an easel pad?

Record the verbatim response (actual words) of the customer/user. Do not make the common mistake of recording the engineering/service solution to his words. For the workshop, brainstorm 20 customer statements.

Figure 4-2

I = inferred E = explicit		CONTEXT OF APPLICATION			
WHO	WHAT	WHERE	WHEN	WHY	HOW
E instructor	E work shop	E bldg 255	for 5 days	historic record	
E trainees	lecture		moved several times	limited work	
I tall		easy to clean walls		E space for all	
I short				to see	

Figure 4-3

Context of Application

A better understanding of the meaning behind customer statements is possible by witnessing the environment, the *Gemba*. Go to the *Gemba*! See it. Feel it. Smell it. Become part of it! Go native!

Who?

Whom do you see using the product? Who may use the product? Who may use the product in the future? The dimensions of the Who column depend upon what may influence the design. The height of the user may be relevant for visibility. (This factor had been ignored in the design for displays in a continuous flow printing press.) Left- or right-handedness may be an issue for hand tools. Pay careful attention to detail. Keep asking the question, "What is missing? What do I need to understand in order to design a world-class product?"

The information in Figure 4-3 may have come from the customer or from observing the customer using the product. An "E" is placed in the narrow column to indicate that the information has come from an external source; that is, it comes from the customer either directly or by observation. Information may also be inferred (generated by the team). An "I" is placed in the narrow column to indicate that information about Who has come from an internal source rather than from the customer.

What?

What are the product's uses? What else might the product be used for, now and in the future? What do you see as uses for the product? In this case, the consultant is using the sheets from the easel pad for the workshops. The sheet is first a work pad. Then it is placed on a wall for ready reference and for presenting a flow of the analysis from one sheet to another.

Where?

Where do you see the product being used? Where else might the product be used, now and in the future? In this case the building and the type of wall surface (easy to clean) define Where. Where could include atmospheric conditions, such as high humidity. Papers curl and cellular phones perform poorly in high humidity.

When?

When do you see the product being used? When may the product be used, now and in the future? In this case, five days represents the maximum time the product will be used. When could refer to frequency, such as continuously or sporadically. When may also refer to the seasons of the year. In one example, five plastic trash cans with a lifetime warranty all failed one day after installation. When was winter with a temperature of -30°C. Where was outside. What was to hold feed for animals.

Why?

Why is this product now selected? Why might it be selected? Why is not the same as What. For instance, the plastic trash can was selected because it opens quietly and workers can remove grain from it with a minimum of noise. The plastic would not rot or rust in the outdoors. In our case, the easel pad solves the problem of limited work space and provides a hard copy of work already done. A white board with the capability to duplicate what was written on the surface is an alternative solution for these two Whys.

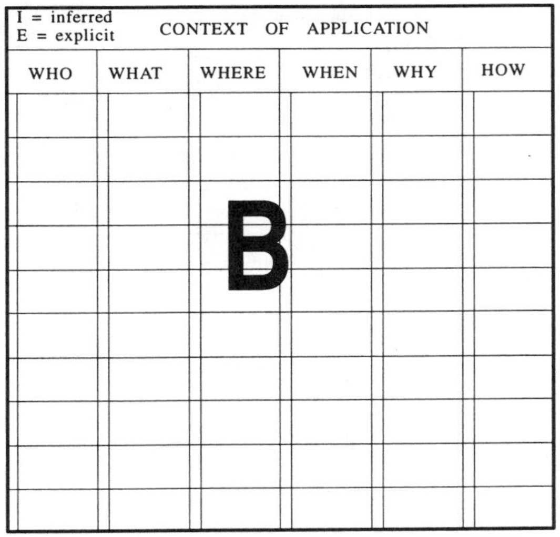

I = inferred E = explicit	CONTEXT OF APPLICATION				
WHO	WHAT	WHERE	WHEN	WHY	HOW

B

Workshop 4-2

Use worksheet B in the back of this manual.

Complete the context table for your application.

Do not look at the verbatim data in the previous workshop but consider the customer segment represented by the customer. In the case example, the consultant is the customer but others will use the product.

The context is for the general case and not for each verbatim response.

Figure 4-4

VOICE OF CUSTOMER TABLE

INFO ABOUT PERSON	VOICE OF CUSTOMER	I = inferred E = explicit	CONTEXT OF APPLICATION					INTEGRATED DATA
		WHO	WHAT	WHERE	WHEN	WHY	HOW	
53 years	I can move it	E instructor	E work shop	E bldg 255	for 5 days	historic record		stays up long time
Consultant	works on my walls	E trainees	lecture		moved several times	limited		sticks to coated walls
1.9 m tall		I tall		easy to clean walls		work space for all	E	many moves possible
12345A		I short				to see		Stays on wall 48 hours
								Repositionable 4 times no change in properties

Figure 4-5

How?

How do you see the product being used? How else might the product be used, now and in the future? Consumers use products creatively. A scented skin lotion has now become a popular insect repellent. Lawn mowers have been used to trim hedges. The customer's unorthodox use of one particular mower resulted in a physical injury lawsuit. The manufacturer lost. In our case, the sheets are used backwards for placing the repositionable note on the surface. The front surface has a coating from which the repositionable notes quickly fall.

Organizations need to interview and observe only 15 to 20 customers in depth.

Research shows that the added benefits of interviewing or observing more than 20 customers are marginal.

Complete Workshop 4-2 as described in Figure 4-4.

Expanding the Verbatim Response

Integrating the context of application with the customer's verbatim response generates an expanded list of information about the customer. In Figure 4-5, the highlighted "works on my walls," when combined with the product being used on "easy to clean walls" and the maximum of "for 5 days," becomes "stays up a long time." This demand also creates "sticks to coated walls." The purpose of this activity is to make sure nothing is missing. It provides one way to fill in the blank spaces of our mosaic picture of the customer. The integrated data should only have one concept in each phrase. "Sticks to coated walls for a long time" represents two ideas, two requirements.

Workshop 4-3 practices the integration of the context and the verbatim data.

This combined document is the Voice of the Customer Table (VOCT) for one customer.

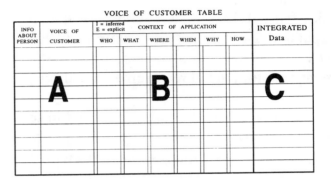

Workshop 4-3

Use worksheet C from the back of the manual.

Place worksheet A to the left of B and worksheet C to the right of B.

Integrate the context with the verbatim response list.

Generate 20 new statements for the customer needs.

Figure 4-6

Sorting Customer Data

The combined list of the original verbatim data and the expanded list will have different types of statements. The expanded list will contain recommended solutions, subjective performance statements, failure modes, specifications, functional requirements and price.

Each of these types of data must be treated in separate analyses. The problems created by mixing different types of data in a single analysis have been the major cause of confusing QFD applications in North America during the past ten years.

This manual looks at the subjective performance requirements and the development of concepts. Deploying the voice of the customer into manufacturing processing conditions is also covered in this manual.

Some definitions of terms used in the QFD process:

FUNCTION: what a product does or what task is performed. Grammatically, it has an active verb plus an object. Active verbs are words such as hold, support, conduct, project and adhere. Vague words such as provide and offer are not useful. A flashlight "providing light" is not as useful for design as a flashlight that "focuses light" or "energizes a bulb." In a function, the product is the implied subject. Accordingly, a statement is not a function if its subject is other than the product.

DEMANDED QUALITY: the customer's subjective description of the performance of the product and its functions. Grammatically, it has an adverb and verb. In English, an adverb often has an "ly" ending. Words like quickly, quietly, smoothly and efficiently imply a fuzzy level of performance. Numbers are typically specifications. The driving demanded quality must be found by asking "Why?" Adding different adverbs to the product's functions facilitates breakthroughs. Demanded qualities should always be worded as positive statements for ease in subsequent analysis.

Some organizations stop after clarifying all the customer's demanded qualities. Prioritized demanded qualities are one input for calculating the importance of the performance measures used in design evaluation.

PERFORMANCE MEASURE: a technical measurement evaluating the performance of some demanded quality. It often states how or what you measure. A performance measure also measures the quality of a function. A function for a flashlight is "illuminates things." One performance measure is the "lumens at 30 meters."

FAILURE MODE: a type of defect. Sometimes a failure mode becomes evident in different environments. Sometimes the defect surfaces after some length of time. Failure modes are often negative statements.

RELIABILITY: the amount of time that passes before a particular failure occurs.

S PECIFICATION: a required numeric value for the product's performance.

TARGET: a desired numeric value for a performance measure.

SOLUTION: a specific design, technology, methodology, manufacturing process or material to be used.

Depending on your particular industry, your designs may have other aspects that require special analysis, such as weight or safety.

A more comprehensive list for the easel pad will be sorted into these different categories (Figure 4-7).

The data from the customer must be sorted into different categories which relate to different aspects of the design process.

Where Does Price Belong?

Price is not a measurement of performance; therefore it will not be used to define design requirements. Chapter 7 will use cost to influence the selection of technologies and design concepts.

There are two ways to include price in the design process. For organizations motivated by the stockholders and next quarter's profits, the selling price is defined by the management's required profit which is added to the cost of the product:

Price = Cost to Manufacturer + Profit

For an organization motivated by increasing market share and making a profit, the target cost for the design is equal to appropriate selling price minus profit desired:

Cost = Price - Profit

Customer Data

A. No bleed-through
B. Common markers work
C. Smear-free
D. All surfaces writable
E. Portable
F. Fits standard easels
G. Large enough to see printing in most meeting rooms
H. Freestanding
I. Stays up five days
J. Sticks to coated walls
K. Sheets remove cleanly with minimum effort
L. Repositionable four times with no change in properties
M. Nondamaging to surfaces
N. Sheets remove with no detectable adhesive transfer
O. Sheets flip
P. Sheet can be removed and posted after previous sheets flipped
Q. Clean edges when torn off
R. Opaque sheet
S. White or off-white
T. I can move it
U. Stays up long time
V. Works on my walls
W. Legible writing medium

Figure 4-7

Teams may use one-word demanded qualities for convenience, but they must have a glossary to explain a statement such as E in Figure 4-7. For this team, "Portable" means "easily transported." The necessity of a glossary cannot be overemphasized. After any discussion about the interpretation of some statement, the agreement must be recorded in a glossary. Teams frequently change the definition during some of the analysis. When this happens, the team can obscure the issue and inadvertently lose sight of the test or measure they are supposed to be discussing. At least one person must refer to the glossary.

Each integrated datum is labeled. If any statement is not a demanded quality, the related demanded qualities are identified.

A. No bleed-through = failure mode. Could result from having a demanded quality of a clean second page after writing on the top page.
  Demanded quality = clean second page

B. Common markers work = demanded quality. This demanded quality implies that many or all of the common markers will leave a mark that is acceptable.

C. Smear-free = demanded quality.
D. All surfaces writable = demanded quality.
E. Portable = demanded quality.
F. Fits standard easels = demanded quality.
G. Large enough to see printing in most meeting rooms = demanded quality.
  G1. Large enough = demanded quality.
  G2. Writing stands out = demanded quality.
H. Freestanding = demanded quality.
I. Stays up five days = specification.
  Demanded quality = stays up long time.

J. Sticks to coated walls = environmental condition.
  Demanded quality = stays on wall.

K. Sheets remove cleanly with minimum effort. Depending upon one's interpretation, this statement could be one concept as presented, or two concepts as shown immediately below. As one concept, the phrase means that no special motions or speed is necessary for the sheet to be removed from the pad. The two concepts would emerge more clearly if the customer had used the word *and* instead of *with*, and this is apt to occur in normal conversation.
  K1. Sheets remove cleanly = demanded quality.
  K2. Easy removal from pad = demanded quality.

Since apparently simple statements may yield more than one demanded quality, it is important to have some identifier associated with each customer who provided data. The team could go back and ask why the customer voiced that demand. Ultimately, the team must get back to the root demanded quality motivator for every customer statement.

L. Repositionable four times with no change in properties = specification. Usually numbers in a statement represent a specification. It is not your customer's responsibility to establish specific requirements, but it is crucial that customers clearly state their demanded qualities. Unfortunately, many customers have had limited choices as consumers and have little experience evaluating products. Going back to the customer's words and the context table prevents this step from becoming a guessing game. Scrutinizing data in this way reduces the chance of misinterpreting customer needs.
  Demanded quality = can be moved many times.

M. Nondamaging to surfaces = failure mode. The demanded quality *removes cleanly from surfaces* is an example of a positive statement of a failure mode.
  Demanded quality = removes cleanly from surfaces.

N. Sheets remove with no detectable adhesive
transfer = failure mode.
Demanded quality = sheet comes off easily.

O. Sheets flip (vague).
Demanded quality = sheet flips on pad.

P. Sheet can be removed and posted after previous
sheets flipped.
Demanded quality = any sheet can be removed.

Q. Clean edges when torn off = demanded quality.

R. Opaque sheet = specifications.
R1. Hides second sheet = demanded quality.
R2. Hides surface = demanded quality.

S. White or off-white = solution.
Demanded quality (same as G2).

T. I can move it (same as E).
Demanded quality = can be moved many times.

U. Stays up long time (same as I).

V. Works on my walls (same as J).

W. Legible writing medium = demanded quality.

While readers may disagree with some of these judgments, they would have to be present and participate in the group discussion before they could justifiably object. Furthermore, not everyone attends all the meetings. Explaining how the team came to every decision is not productive. For this reason, in addition to creating a glossary, the project must document important discussions. Documentation must be sufficient to answer any thoughtful question that someone who was not present might ask. The need for documentation is especially important for a future team that will use the information for a model upgrade. It

is unlikely that a future team will have the same team members. Even if the whole team is reassembled, people's memories seem to change the facts after several months.

Some organizations find that they have learned enough after this analysis and that continuing the QFD process is not necessary. Such a decision can occur after a major breakthrough, or when the team feels that they do not have the resources for any additional breakthroughs.

It is time to and review your understanding.

Workshop 4-4

Identify the types of statements you have.

Determine the likely demanded qualities for any statements that are not demanded qualities.

Record all demanded qualities on index card size removable notes.

Print in block letters using a black felt tip water-base marker.

Fellow team members must be able to read the notes from a distance of 3 meters in a later workshop.

Figure 4-8

Affinity Diagrams

The affinity diagram provides structure for verbal data by creating natural clusters or groups. The groupings increase understanding in the same way a bar chart can change hundreds of numbers from data into information.

These tools ensure that the list of demanded qualities is complete and expressed at the same level of detail. The affinity diagram defines natural groups of demanded qualities and helps identify any missing demands. The affinity diagram is only one step in the Jiro Kawakita (KJ) Method. The KJ Method is much more involved because it was originally developed to structure the textual observations of the cultural anthropologist Jiro Kawakita. The tree, as described on page 70, can verify the affinity diagram's structure and completeness. The tree assures that the same level of abstraction is used in questionnaires and later analysis. Eventually, demanded qualities with the same level of abstraction will be used in a questionnaire with a larger group of customers. The questionnaire identifies the importance of each of the demanded qualities. Teams also gather the customers' evaluations of both their own and the competition's existing products.

Each demanded quality is placed on a 3x5 card (about 7.5 cm x 12.5 cm) or a repositionable note, like those you have already made for Workshop 4-4.

The affinity process resembles a primary school mental exercise. The children look at several shapes and group them by answering the question, "How are these things the same; how are they alike?"

The next part of the process is structured to change your usual paradigm of product understanding. The following example illustrates clearly the activity's purpose.

Suppose you have just won a large sum of money from a lottery. You and your spouse have decided to hire an architect to design your dream home for your dream location. You interview two architects. The first architect asks you what style you want. "Colonial!" is your answer. The first architect continues with more questions:

1. How large is your family?

2. What are your family members' ages?

3. How large do you expect the family to become?

4. How much entertaining do you do?

5. What hobbies do you enjoy?

The architect states he can complete the design in one month.

The second architect asks you to describe your dream location and why it is so special to you. Then she asks you to invite your close friends, relatives and anyone else who may enjoy this structure to a meeting. During the meeting the total group brainstorms the activities that will occur in your home. Next the list is recorded and placed on 3x5 cards. Your guests then find natural clusters of activities and group them together; they do this in silence. This seems strange, but you follow the directions because of the architect's reputation for delighted clients. The architect announces that she will use the group's information to design the space to support these activities and enhance your location.

Which home would you enjoy the most?

Which home could you easily resell?

The first architect had pigeonholes for each of your answers. These were conventional spaces, such as the bedroom, the kitchen, etc. The number of items that landed in each pigeonhole would dictate how many square meters he would devote to each of the conventionally defined spaces.

The second architect took a more organic and creative approach. Your particular needs created the space. At the same time, the second architect's design would be more difficult to sell. If you had no intention of ever selling the home, the second architect would be best. However, if your job or lifestyle suggests that you would not keep the house for very long, you should choose the first architect.

If you, your friends and relatives talked during this exercise, your procedure would resemble the first architect's approach; you would probably identify conventional clusters before grouping your demanded qualities. Looking at the demanded qualities with a fresh perspective facilitates creative breakthroughs. One way to do this is to "close down" the left side of the brain, so that the right, creative side can take charge. Research into many cultures shows that the left brain is logical and structured, while the right brain is creative and holistic. Do you think of creative solutions to your work problems while driving home? Driving is a right-brain activity; thus most of your thinking as you drive occurs in the right brain rather than the left. Grouping the cards with your nondominant hand also facilitates right-brain activity.

AFFINITY DIAGRAM

C. Smear-free
B. Common markers work
A. Clean second page
G2. Writing stands out
W. Legible writing medium

D. All surfaces writable
G1. Large enough

E. Portable
F. Fits standard easels
H. Freestanding

K2. Easy removal from pad
O. Sheet flips on pad
P. Any sheet can be removed

R1. Hides second sheet
R2. Hides surface

J. Stays on wall
L. Can be moved many times
N. Comes off easily

Figure 4-9

The process begins by placing all the cards on a table randomly, or by placing repositionable notes on a vertical surface. If the team has not seen the complete list, spend a moment reading all of the demanded qualities. Then have everyone start at the same time. When all participants are ready, everyone **quickly and without thought** finds two demanded qualities that have something in common. Continue to add all related demanded qualities to this group of two. If you find a demanded quality is not where you think it belongs, move it. If it is moved again, make a duplicate and talk about it later. The process continues until all the demanded qualities are in a group. A group can contain only one demanded quality. The results of this process for the easel pad are shown in Figure 4-9. As a guideline, a group with more than 7 ± 2 members may become two or more groups. Nine group members is the limit for the prioritization process to be shown later.

Single-item groups may not be demanded qualities, or they may be of a unique nature requiring special treatment during the design.

Workshop 4-5a

Follow the process described for the easel pad to form the natural clusters for all of your demanded qualities.

Group the notes quickly, without talking. Use your nondominant hand.

Figure 4-10

Begin discussion after group composition for the demanded qualities becomes stable. First review those demanded qualities that seemed to have more than one home. This situation can occur if team members have different interpretations of a demanded quality's meaning. It could also indicate that the team should clearly define the commonality for some of these groups.

Next, select a descriptive name for the group. A dominant team member can talk everyone into using traditional groupings. Be careful! Take advantage of fresh insights offered by the new arrangement of information. Group names must also be demanded qualities, but at a higher level of abstraction (Figure 4-11).

This procedure is most effective when customers do the grouping.

The natural clusters that form within the demanded qualities provide more insight than structured questionnaires. While questionnaires ask if the customer wants a particular feature, brainstorming and the affinity diagram allow the customers to create the list and its structure themselves.

So far, the process has ignored the demanded quality's level of abstraction. Look at each group and judge if all elements are at the same level of abstraction. In Figure 4-9 we see that Legible Writing Medium would include all the other demanded qualities. Legible Writing Medium can become the name of that group. This group heading is highlighted in Figure 4-11. Since all the members of the groups are demanded qualities, the group names must be demanded qualities of a more general nature.

AFFINITY DIAGRAM

LEGIBLE WRITING MEDIUM

C. Smear-free
B. Common markers work
A. Clean second page
G2. Writing stands out

MANY WAYS TO STAND

E. Portable
F. Fits standard easels
H. Freestanding

INDEPENDENT SHEET MANIPULATION

K2. Easy removal from pad
O. Sheet flips on pad
P. Any sheet can be removed

RAPID SHEET POSITIONING ON/OFF WALL

J. Stays on wall
L. Can be moved many times
N. Comes off easily

USABLE IN MANY DIFFERENT SPACES

D. All surfaces writable
G1. Large enough

FREE OF SUBSURFACE INFORMATION

R1. Hides second sheet
R2. Hides surface

Figure 4-11

AFFINITY DIAGRAM

LEGIBLE WRITING MEDIUM

C. Smear-free
B. Common markers work
A. Clean second page
G2. Writing stands out

USABLE IN MANY
DIFFERENT SPACES

D. All surfaces writable
G1. Large enough

MANY WAYS TO STAND

E. Portable
F. Fits standard easels
H. Freestanding

CONVENIENT PACKAGE

Protects pads
Opens easily
Pad can be packaged

INDEPENDENT SHEET
MANIPULATION

K2. Easy removal from pad
O. Sheet flips on pad
P. Any sheet can be removed
   Removal is neat

FREE OF SUBSURFACE
INFORMATION

R1. Hides second sheet
R2. Hides surface

RAPID SHEET POSITIONING
ON/OFF WALL

J. Stays on wall
L. Can be moved many times
N. Comes off easily
   Easily positioned

Figure 4-12

Group names should have a minimum of two words, with one of them describing a level of performance. Engineers tend to like one-word group names, such as Writing. Do not take the easy way out; use descriptive words and avoid generic or vague modifiers such as "good." Complete Workshop 4-5b in Figure 4-13.

The next activity is checking each group by asking the question, "If this is the name of the group, what elements should be included but are missing?" There are two examples of added elements in the diagram. Discussing commonalties and the selection of a group title may generate additional members to the groups. For example, if the group heading for a TV controller is labeled Easily Used, it might suggest Older People Can Use It as an added member of the group. Any additional demanded qualities should be added at this time. Consider using the group heading with a tree diagram to help generate a more comprehensive list.

Next test for missing groups by asking the question, "When considering all aspects of the product, are there any aspects not represented by the group headings?" In the example, the group of elements related to packaging was completely ignored (Figure 4-12).

Several years ago, the following story was shared at a QFD public seminar. When an international company was improving its repositionable notes, it used focus groups. These groups were unknowingly evaluating different formulations and competitors' products. To keep the product manufacturers anonymous, the company removed the back sheets containing the company's logo or other identifier and replaced them with colored sheets. The group members then compared red to blue, etc.

Ultimately, the focus group helped select the best formulation, but a significant design flaw was overlooked. Customers who used the product kept a razor blade in their desks to open the package. The tear strip was transparent and only the manufacturer knew where it was. The focus group members never had to open a package because of the backing change and the difficulty of packaging 20 special packs.

Our easel pad team had the same problem, there was no group that had elements for packaging. The group and its elements were subsequently added (Figure 4-12).

Complete Workshop 4-5c in Figure 4-14.

Workshop 4-5b

Follow the process described for the easel pad to select group headings.

Are all of the elements of a group at the same level of abstraction?

If one of the elements is more general, can it be used as a group name?

Figure 4-13

Workshop 4-5c

Continuing with your affinity groups:

Are there any missing groups? Add these groups.

Are there any missing elements in any of the groups?

For practice only, add to one of the former groups and also add one new group.

Figure 4-14

Obtaining competitive assessments is a challenge for many organizations. One method is the sample questionnaire in Figure 4-15 that uses the five-point scale. Some organizations use "influence to buy" rather than importance. In this case, a 1 means the demanded quality is of so little importance that the customers would not change brands if they experience poor performance in this area. A 5 means that any displeasure with the performance will cause the customer to change brands. In this example, the customers being surveyed have ready access to several brands of the product. They evaluate each product using the five-point scale. Do not simply average the data in the Influence column, but look at the distribution. Knowing the various needs of different kinds of customers is essential to QFD. For instance, all sampled consumers selecting an importance of 3 is not the same as half selecting a 1 and half selecting a 5. This consumer rating system is flexible enough to allow adaptation to your organization's unique situation.

Given the way this questionnaire is written, the customers have an opportunity to evaluate several suppliers of this type of product. A 5 in this point scale means the customer believes the product performance is very good, and a 1 is used if performance is very bad.

Sample Questionnaire

IN COLUMN 1 THERE ARE SEVERAL CRITERIA FOR JUDGING THIS PRODUCT. IN COLUMN 2 YOU ARE ASKED TO SELECT HOW MUCH EACH ITEM WILL INFLUENCE YOUR DECISION TO PURCHASE.

WHILE YOU ARE HERE, YOU CAN USE THE DIFFERENT VERSIONS OF THIS PRODUCT.

COLUMN 3 IS USED TO COMPARE THE DIFFERENT VERSIONS.

PLEASE CIRCLE THE CURRENT PRODUCER OF THE PRODUCT YOU OWN. IS IT X, Y or Z?

ITEMS TO JUDGE PRODUCT	INFLUENCE					PERFORMANCE					
	VERY STRONG	STRONG	MILD	LITTLE	DOES NOT MATTER		VERY GOOD	GOOD	SO-SO	BAD	VERY BAD
FREE-STANDING	5	④	3	2	1	X	5	④	3	2	1
						Y	5	4	3	②	1
						Z	⑤	4			1
WRITING STANDS OUT								4	③	2	1

Figure 4-15

Kano Model

The Kano Model is used to understand the importance of functions or features to a customer. These functions or features are called needs. Paired questions describe a level of performance for a function which is a demanded quality. This tool sorts consumer needs into one of three categories: excitement needs, performance needs and basic needs. This information can be gathered using a questionnaire.

Basic Needs

Some of these needs are so fundamental that they are not expressed by the consumer; however, because they are so crucial, they must be identified. You would expect a car to start in cold weather, your toaster to work with different types of bread or all characters in a software package to be printable on every printer allowed. For example, note the circles not present around the faces in the Kano Model in Figure 4-16; changing from a dot matix printer to a laser printer caused this omission. The best performance for a basic need will only result in a consumer who is not unhappy.

Performance Needs

Performance needs provide an increase in satisfaction as performance improves. These kinds of needs are generally expressed by the consumer. One example would be the number of miles per gallon for a car. The better the performance, the happier the customer.

Excitement Needs

Excitement needs cause immediate happiness. Further increases in performance cause more delight. Needs of this type are also not verbalized, possibly because we are seldom asked to express our dreams. Creation of some excitement features in a design would differentiate your product from the competition.

As a product improves, the needs change categories. It does not take long for an excitement need to become a performance need. Similarly, over time the performance need becomes a basic need. Likewise, a drastic improvement in performance can create excitement again.

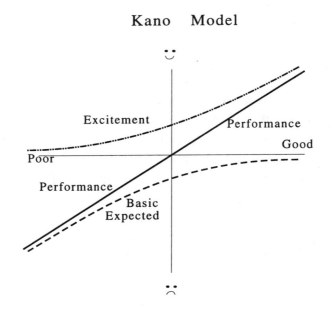

Figure 4-16

The time required to deliver a letter across the continent can serve as a performance need example. A century ago, having a letter reach the opposite coast was in itself an accomplishment (excitement). The Pony Express provided a solution, setting a new standard for the postal service—the shorter the delivery time, the happier the consumer (performance need). Within the last decade, we have come to expect a letter to take only a few days to cross North America (now a basic need). We saw the process repeated when, a few years ago, the FAX machine became an excitement need because of its nearly instant delivery of a letter. Today the FAX machine is being replaced by e-mail and the Internet. Video conferencing is about to explode with the introduction of systems that work with common desktop computers.

Kano's Paired Questions

Kano used a set of paired questions to determine which of the three categories of needs applied to a particular airline passenger. The first question asks how you feel if (something that exists). The second question asks how you feel if (something that does not exist). The following example applies this approach to an instructor's performance.

Use 1-5 to answer: 1. I really like it, 2. I like it, 3. I feel neutral, 4. I do not like it, 5. I really do not like it.

Record the answer to each question after the example answers. Workshop 4-6 requires this information.

1. How do you feel if the instructor has a good sense of humor? (answer: ex. 1, you _____)

2. How do you feel if the instructor presents much useful information? (answer: ex. 3, you _____)

Use A-E to answer: A. I really like it, B. I like it, C. I feel neutral, D. I do not like it, E. I really do not like it.

3. How do you feel if the instructor does not have a good sense of humor? (answer: ex. C, you _____)

4. How do you feel if the instructor does not present much useful information? (answer: ex. E, you _____)

The first and third questions are the paired questions evaluating humor in a presentation. Humor is probably an excitement or performance need. The second and fourth questions are the paired questions evaluating useful information in a seminar. Useful information is probably a basic or performance need.

By Kano's Model (Figure 4-16):

Excitement needs are recorded in cells 1C or 1D.

Performance needs are recorded in cell 1E.

Basic needs are recorded in cells 2E or 3E.

To record the examples in the matrix, select the row to match the number for question 1 and select the column to match the letter for question 3, and place an $\mathcal{H}$ for Humor. Now record your answer.

Select the row to match the number for question 2 and select the column to match the letter for question 4, and place a $\mathcal{C}$ for Content. Record your answer.

		NEGATIVE → I REALLY LIKE IT (A)	I LIKE IT (B)	I FEEL NEUTRAL (C)	I DO NOT LIKE IT (D)	I REALLY DO NOT LIKE IT (E)
POSITIVE		A	B	C	D	E
1	I REALLY LIKE IT			$\mathcal{H}$		
2	I LIKE IT					
3	I FEEL NEUTRAL					$\mathcal{C}$
4	I DO NOT LIKE IT					
5	I REALLY DO NOT LIKE IT	R				

I once gave a seminar for a group in which several of the participants had useful information as an excitement need. This is an indictment of the typical course this group of people had attended.

These two pairs of questions should be administered to future class participants to get a better match between their needs and the instructor. If humor is a basic need, then it would be more productive to hire a comedian than a technical expert.

If cell 5,A has an entry, it represents a reversal (R). The person is unhappy with the presence of something and happy with its absence. For example, the paired questions for restaurant clientele are:

For Sorting Customer Needs

POSITIVE		NEGATIVE → I REALLY LIKE IT (A)	I LIKE IT (B)	I FEEL NEUTRAL (C)	I DO NOT LIKE IT (D)	I REALLY DO NOT LIKE IT (E)
		A	B	C	D	E
1	I REALLY LIKE IT	?	I	E	E	P
2	I LIKE IT	I	?	I	I	B
3	I FEEL NEUTRAL	RE	I	?	I	B
4	I DO NOT LIKE IT	RE	I	I	?	I
5	I REALLY DO NOT LIKE IT	RP	RB	RB	I	?

EXPLANATION
E excitement
P performance
B basic
I indifferent
? inconsistant
RE reversal excitment
RP reversal performance
RB reversal basic

Workshop 4-6

Select the row to match the number for question 1 and select the column to match the letter for question 3, and place an H for Humor.

Select the row to match the number for question 2 and select the column to match the letter for question 4, and place a C for Content.

For you, are humor and content excitement, performance or basic needs?

For a class, use separate matrices for humor and content. Record the number in each cell.

Discuss your results.

Figure 4-17

1. How do you feel if the restaurant has meat on the menu?

2. How do you feel is the restaurant does not have meat on the menu?

A meat-eating individual would feel neutral (3) for question 1 and really not like it (E) for question 2. The combination of answers indicates that getting meat at a restaurant is expected, a basic need, for this individual.

A vegetarian who has been looking for a place to eat for several weeks would really not like it (5) for question 1 and would really like it (A) for question 2 (a reversal). By folding the result along the 1A, 5E axis we see that the vegetarian is excited by not having meat on the menu.

⌐o━━╦ Devoting resources to improving basic needs to make a customer happy is impractical.

It is more productive to devote limited resources to creating excitement features or improving performance needs.

The Tree and Prioritizing Demanded Quality

The *tree* provides a method to verify the levels of abstraction and find missing data. The tree can be constructed in two ways: from the branches to the trunk or from the trunk to the branches.

When going from the trunk to the branches, the starting point is the group headings. This is a good way of identifying missing groups and checking the consistency of abstraction in the group headings. A very large project may have as many as eight levels of detail for the data. Three levels are sufficient for most applications.

Using the results of the affinity diagram is equivalent to starting with the leaves and working toward the trunk.

AHP or the distribution of 100 points can be integrated with the tree structure. Before continuing with the easel pad example, consider a case that illustrates how AHP and the tree can work together.

Suppose the affinity diagram had three groups with three elements all relating to a Friendly Pad.

Friendly Pad

Looks Exciting (LE)	Easy to Use (EU)	Status Enhancer (SE)
LE1	EU1	SE1
LE2	EU2	SE2
LE3	EU3	SE3

This information is quickly converted into a tree.

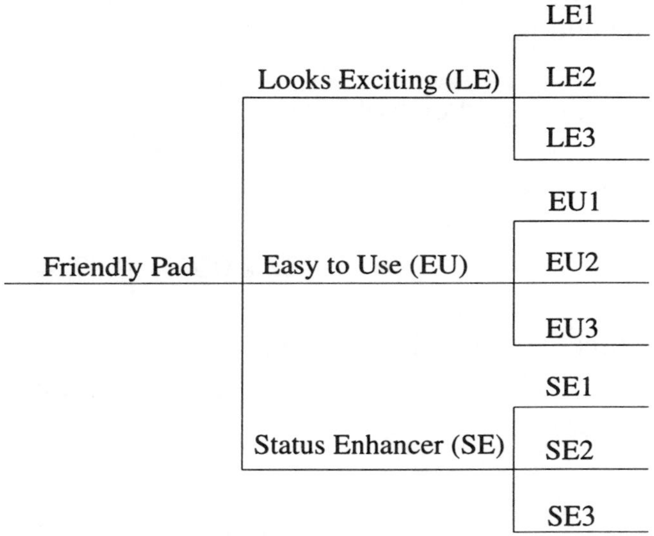

Figure 4-18

The AHP process is first used to compare the relative importance of various group headings.

	LE	EU	SE
LE	1.00	0.33	0.25
EU	3.00	1.00	0.50
SE	4.00	2.00	1.00

Figure 4-19

The initial AHP calculation is shown, which results in an importance of 0.122, 0.320 and 0.558 for Looks Exciting, Easy to Use and Status Enhancer, respectively. This process for calculating the importance is explained in detail in Chapter 3. The importance information is placed on the tree (Figure 4-20).

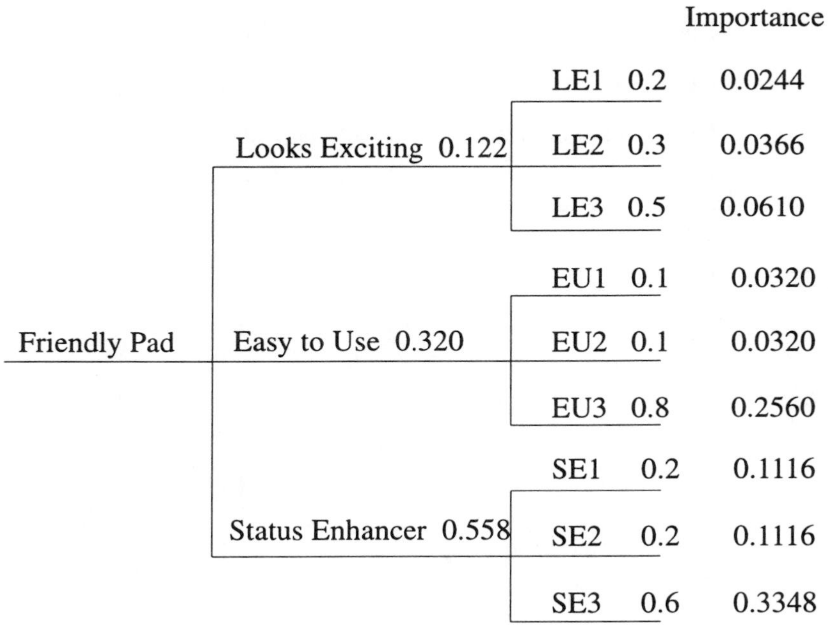

Figure 4-20

Next the relative importance of the elements for Looks Exciting is determined and recorded on the tree. This process is continued for each of the other groups.

PRIORITY OF DEMANDED QUALITY	Smear-Free	Common Markers	Clean Second Page	Writing Stands Out
Smear-Free	1	.2	3	.33
Common Markers	5	1	7	3
Clean Second Page	.33	.14	1	.2
Writing Stands Out	3	.33	5	1
Total	9.33	1.68	16	4.53

	Smear-Free	Common Markers	Clean Second Page	Writing Stands Out	Row Avg	Legible Avg	Import.
Smear-Free	.11	.12	.19	.07	.12	.34	.04
Common Markers	.54	.60	.44	.66	.56	.34	.19
Clean Second Page	.04	.09	.06	.04	.06	.34	.02
Writing Stands Out	.32	.20	.31	.22	.26	.34	.09
Total	1.0	1.0	1.0	1.0	1.0	-	-

Figure 4-21

The overall importance is the product of a group heading and its elements.

The overall importance of LE1 is:

$$\text{Importance of LE1} = \text{LE} \times \text{LE1}$$

$$= 0.122 \times 0.2 = 0.0224$$

Using this structure makes a comparison of the importance of any lower level demanded qualities meaningful.

This same approach is possible with the distribution of 100 points. The 100 points would be distributed among the group headings. If 20 points were allocated to Looks Exciting, then the 20 points would be allocated among the three elements of this group. The process would be continued for the other groups. A comparison of the points for the lower level demanded qualities is meaningful.

For larger projects, repeat the process for all levels of detail.

Remember that Saaty recommends a maximum of nine branches for any comparison. If you have more than nine groups, break the data down into more levels.

This process was used for the easel pad (Figure 4-21). The calculation for relative importance within the group Legible Writing Medium and overall importance is shown in Figure 4-22.

A tree structure can be adapted to spread sheet analysis, as shown for all the demanded qualities of the easel pad.

🔑 This prioritization information should come from the customer, not the design team.

There are always exceptions, but be warned that using numbers to represent guesses of what the customer feels and believes often leads to flawed designs.

When working with a group of customers, the group consensus is used for the cell entries. If you must find the average, the geometric mean is the appropriate calculation. To find the geometric mean of 5 numbers, find the product and then the 5th root. This method is necessary, as shown in the following example. Two customers differ: one feels that X is twice as important as Y, but the second believes that X is 1/2 as important as Y. The average should be equally important. The arithmetic mean, however, shows that X is 1.25 as important a Y. The geometric mean shows them as equally important. Again, be careful of different markets and look at the distribution.

PRIMARY		SECONDARY		Imp.
LEGIBLE WRITING MEDIUM	.34	Smear-free	.12	.04
		Common markers work	.56	.19
		Clean second page	.06	.02
		Writing stands out	.26	.09
MANY WAYS TO STAND	.07	Portable	.12	.01
		Fits standard easels	.55	.04
		Freestanding	.33	.02
INDEPENDENT SHEET MANIPULATION	.21	Easy removal from pad	.39	.08
		Sheet flips on pad	.35	.08
		Any sheet can be removed	.19	.04
		Removal is neat	.08	.02
RAPID POSITIONING ON/OFF WALL	.20	Stays on wall	.59	.12
		Can be moved many times	.06	.01
		Comes off cleanly	.20	.04
		Comes off easily	.15	.03
USABLE IN MANY DIFFERENT SPACES	.04	All surfaces writable	.75	.03
		Large enough	.25	.01
CONVENIENT PACKAGE	.03	Opens easily	.22	.01
		Protects pads	.73	.02
		Pad can be packaged	.05	.00
FREE OF SUBSURFACE INFORMATION	.11	Hides second sheet	.69	.08
		Hides surface	.31	.03

Figure 4-22

Workshop 4-7

Take the affinity diagram data and place it in a tree structure.

Distribute 100 points to determine the importance of your groups and their elements.

Calculate the overall importance of all your elements.

Figure 4-23

The two most important demanded qualities are Common Markers Work and Stays On Wall. According to the results, 19% of the overall value of the easel pad is its ability to accept common markers. Similarly, 12% of the value is its ability to stick to the wall. For some organizations, this prioritization is sufficient and they choose to continue with their traditional design process.

It is time to and review your understanding.

If your team is looking for added benefits from the QFD process, you now have sufficient information about the customer to begin the translation of the customer's requirements into the technical requirements for your design process.

Keep on stepping!

Chapter 5

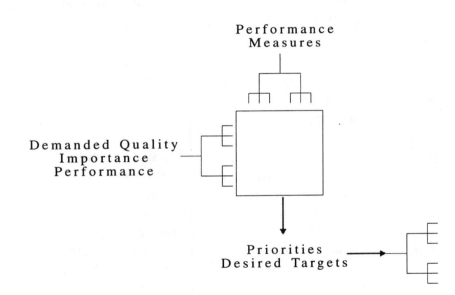

Performance
Measures

Demanded Quality
Importance
Performance

Priorities
Desired Targets

Customer Voice into Design Team Voice

Upon completing this chapter, you will be able to:

♦ Calculate the composite importance for the customer's demanded quality.

♦ Calculate the priorities of design requirements.

♦ Establish meaningful design targets.

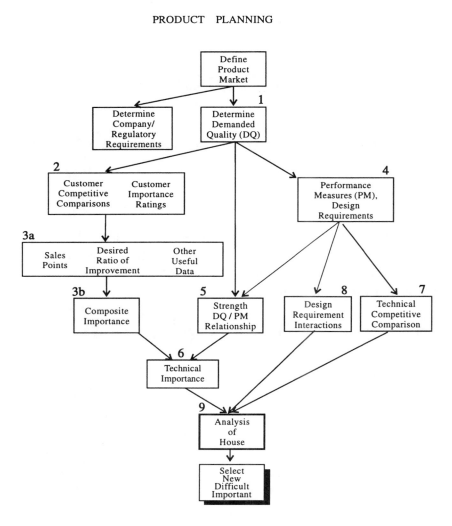

PRODUCT PLANNING

Figure 5-1

Overview

The first step of product planning is to make visible the subjective demanded qualities voiced by the customer. These are translated into the engineering terms for the performance measures used by design organizations.

Several steps are necessary to accomplish this translation. Better translations make it more likely that the product will be world class in the eyes of the customer.

The flow chart in Figure 5-1 depicts the sequence of nine steps. Parallel activities, as shown in the flow chart, can reduce the total project time.

The process starts by defining your product market correctly. Who is the customer? How do you determine the desires/needs of the customer? Are there federal regulations controlling the performance of the product? Is there any particular aspect of the product that is defined by your organization's image? For example, Harley Davidson stopped a QFD design project because the customers' desires for a particular segment were not appropriate for a Harley Davidson motor cycle. This decision made Harley Davidson avoid the disaster of stopping the project after costly development.

Chapter 4 sorted and expanded the customer's/user's verbal information. This information was then sorted, by customer, into groups representing different aspects of a design process. Some of these were the customer-demanded qualities, the functions of the product and the failure modes of the product. Only the demanded quality will be used for this phase of step-by-step QFD.

In Figure 5-1, this list of the Demanded Qualities (1) of the consumer/customer is used to gather the Customer Importance Ratings (2) and subjective Customer Competitive Comparisons (2). This trilogy becomes the driving force in the remaking of the demanded qualities into performance measures. The performance measures are the language your organization uses to evaluate alternative designs.

Design teams use this customer information to clarify the Sales Points (3a) of marketing strategies and the planned improvement in satisfaction desired for each of the customer-demanded qualities (the Ratio of Improvement (3a)). Other Useful Data (3a), such as service complaints, support these decisions.

Composite Importance (3b) is the product of Customer Importance (2), Sales Points (3a) and the amount of effort necessary for design (Ratio of Improvement (3a)). The composite importance is calculated for every demanded quality.

At least one objective way to measure the level of performance is identified for each demanded quality (Performance Measures (4)). These performance measures are also called the design requirements, the quality characteristics and the substitute quality characteristics. This manual usually uses the term

performance measure to emphasize the measurement rather than the solution for a design.

The degree to which a performance measure predicts the satisfaction of the consumer for each demanded quality (Strength of DQ/PM Relationship (5)) is the link between the Performance Measures (4) and Composite Importance (3b) used to calculate the Technical Importance (6).

For projects working on a model upgrade for an existing design, it is possible to look for conflicts between performance measures (Design Requirement Interactions (8)). The identification of conflicts between performance measures this early in a design project provides more time to design out the conflicts, or at least to make some conscious compromises.

A technical comparison of the performance of the product against the competition (Technical Competitive Comparison (7)) is necessary to define the targets for the critical performance measures.

All of the information depicted in the flow chart can be presented in a single graphic document. The picture is a composite of a matrix and several tables. If the interaction information between performance measures is included, the form is often called a **House of Quality**. This final document is shown at the end of this chapter (Figure 5-25). Each one of the steps in the flow chart matches activities in your normal design process and will be presented in this format.

DEMANDED
QUALITY

| Smear-Free |
| Common Markers |
| |
| Freestanding |
| |
| Stays on Wall |
| |
| Easily Removed from Wall |
| |
| Opens Easily |
| Protects |

Figure 5-2

Demanded Quality

Every product design process mentions product design requirements. The first step is to list all the demanded qualities at the same level of abstraction. In the tree in Chapter 4 there are seven demanded qualities at the primary level and 23 at the secondary level. Level selection is one way to control the analysis effort. If the team used 100 demanded qualities, then a matrix of dimension at least 100 by 100 would require discussing 10,000 relationships. Selecting too general a level would result in a very small matrix of limited value. As a rule of thumb, 20 to 30 demanded qualities are comfortable to analyze and provide useful insight.

For the purposes of the easel pad example, seven demanded qualities were selected from five of the seven affinity groups. The entries are highlighted in the table with the tree format in Chapter 4, Figure 4-22. The blank rows in Figure 5-2 are used to separate the groups.

Which demanded qualities should an organization select? For every rule there is an exception, and so our standard answer is "it depends." In this case, which demanded qualities an organization selects depends upon what the organization is interested in learning and how much direction is needed. A microwave popcorn project had 70 demanded qualities at the secondary level or 10 at the primary level. The 70 were selected because the 10 primary demanded qualities would have yielded a very coarse analysis. For another QFD study, a very small matrix was used to select which system of an automobile to use for a detailed QFD project. The small matrix was driven by general customer requirements.

Another way to reduce the size of the analysis is to select only the more important demanded qualities. In the case of the easel pad application, two of the demanded qualities, Accepts Common Markers (19%) and Stays on Wall (12%), represent 31% of the requested demanded quality. This could be enough for a project team and the last step in this application of step-by-step QFD.

The Kano Model, introduced in Chapter 1, suggests that basic needs should not be part of the analysis because they must be satisfied and are all extremely important. However, they should be listed and carried along with the excitement- and performance-related needs.

It is time to

and review your understanding.

Some organizations find that working to improve the most important demanded qualities is all they need from QFD and they return to their traditional design process.

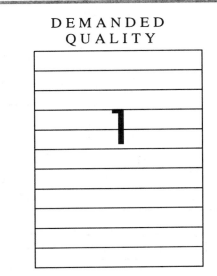

Workshop 5-1

Nine pieces of the QFD puzzle are included with this workbook. One is a triangle with the number eight. The other eight are rectangles with the numbers 1 through 9, excluding 8. Use piece 1 for this workshop. All the pieces for an individual user are in the Worksheets section of this manual.

Your task is to select eight demanded qualities from the workshops in Chapter 4. You may also generate a new list. In either case, use the second level of detail and select two or three demanded qualities from three of the affinity groups.

Figure 5-3

DEMANDED QUALITY	A Importance	B XYZ	C Them 1	D Them 2	E Them 3
			Customer Evaluation		
Smear-Free	4	4	5	5	5
Common Markers	19	4	5	5	3
Freestanding	2	5	2	3	1
Stays on Wall	12	3	4	2	1
Easily Removed from Wall	3	4	3	2	2
Opens Easily	1	2	3	3	5
Protects	2	2	3	2	5

Figure 5-4

Quality Planning Table (part 1)

Sometimes during the design process the team discusses how well the competition is doing and the importance of different product characteristics for the customer. Gathering this information from the customer is more realistic.

The Quality Planning Table begins with the customer input of Demanded Quality, Importance for each demanded quality, and the subjective evaluation of product performance for several competitors. This information is used to establish a Composite Importance for each demanded quality. The composite is the product of the Importance, Ratio of Improvement and Sales Points.

In Column A, Importance comes from the customer through the affinity diagram and tree in Chapter 4. These entries are written as percentages for visibility and are preferred by some organizations for communication. Use whichever form you prefer.

Many organizations still use the following five-point scale for expressing the Importance:

5 demanded quality is a primary influence upon the decision to buy

1 demanded quality has no influence upon the decision to buy

Marketing can gather customer opinions of your product (XYZ, column B) and of different competitors' products (Them, columns C, D and E). However, QFD requires identification of the demanded qualities rather than demographic information, often provided by marketing.

QFD teams find it very productive for all members to meet with and observe a focus group composed of a few users of the product. All too often, organizations skip this important step. There are some organizations that have not allowed the competitor's product in their facilities. Some QFD teams have members who have never talked to a real customer. There are also teams that have never experienced the environment where the product is used. The end result is a design team that is unaware of what their customers really want. My favorite quote from designers who are out of touch with the real user/customer is, "They cannot possibly want that. It doesn't make sense."

Organizations can derive great benefits from focus groups. The following anecdote illustrates the perils of using these groups without appropriate organizational support. Company ABC (the name has been changed to protect feelings and reputations) designs and manufactures office furniture. The president asked an internationally known designer to create the next model of a particular product line. The president approved the new concept and a mockup was made for the focus groups. The focus groups were video taped. They strongly disliked the new furniture. The design team was uncomfortable telling the president the outcome of the focus groups, because the consultant was expensive and a personal friend of the president. They suggested the president watch the video tape. Afterwards, he stated that the company would never use focus groups again. "What did they know?!" It is possible the president was a visionary or that he believed that the members of the focus groups did not represent the significant stakeholders. What do you think?

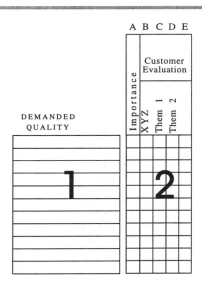

Workshop 5-2

Take piece 2 and place it to the right of piece 1, as shown above. Your task is to take the importance values from Chapter 4 or guess real customer-demanded-quality information. Remember, this is a simulation and is not a recommended way to gather Importance information.

For each product, record customer performance ratings for each demanded quality using the five-point scale defined in Chapter 4. All 1's or 5's are possible for a given demanded quality. AHP could be used for this process, but not for this particular workshop.

Figure 5-5

Once organizations have used focus groups to compare their products with the competition, they can use a graph to show all of the product evaluations and provide a good overall summary of their products and the competition (Figure 5-9).

Quality Planning Table (part 2)

The traditional design process must choose directions for improvement and decide what aspects of the product will be used to promote sales. Part two of the Quality Planning Table provides a structure for these discussions and a means to display the summary of these decisions.

The first demanded quality, Smear-Free, is not very important; a 4 compared to the 19 for Common Markers (column A). The current satisfaction is a 4 compared to 5's for all the competitors (column B). Trade magazines and others do comparative reporting which is helpful.

The target performance (column G) is selected for each demanded quality. Target is influenced by the organization's performance in relation to competitors' and the customer's demanded quality Importance. It also considers the company's strategic plan and competitors' development plans. It would be reasonable not to consider improving this demanded quality, but the team decided on a target of 5. The Ratio of Improvement is equal to the Target divided by the current judgment for XYZ. Using the column headings:

$$H = G / B$$
$$= 5 / 4 = 1.25$$

This is rounded up in the table to 1.3 for presentation.

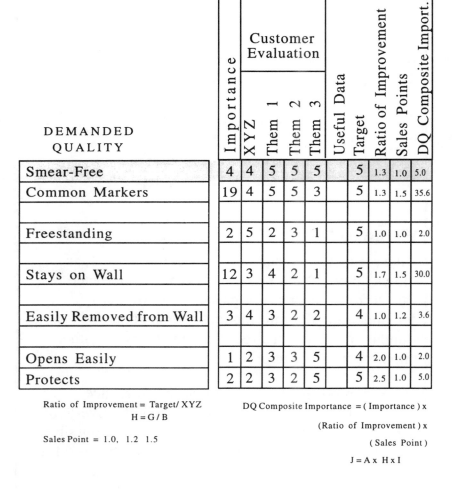

	A	B	C	D	E	F	G	H	I	J
DEMANDED QUALITY	Importance	XYZ	Them 1	Them 2	Them 3	Useful Data	Target	Ratio of Improvement	Sales Points	DQ Composite Import.
Smear-Free	4	4	5	5	5		5	1.3	1.0	5.0
Common Markers	19	4	5	5	3		5	1.3	1.5	35.6
Freestanding	2	5	2	3	1		5	1.0	1.0	2.0
Stays on Wall	12	3	4	2	1		5	1.7	1.5	30.0
Easily Removed from Wall	3	4	3	2	2		4	1.0	1.2	3.6
Opens Easily	1	2	3	3	5		4	2.0	1.0	2.0
Protects	2	2	3	2	5		5	2.5	1.0	5.0

Ratio of Improvement = Target/ XYZ
H = G / B

Sales Point = 1.0, 1.2 1.5

DQ Composite Importance = (Importance) x

(Ratio of Improvement) x

(Sales Point)

J = A x H x I

Figure 5-6

Sales Points are the specific features that will distinguish a product from the competition. As Workshop 5-3 indicates, trying to improve every aspect of a product is not an efficient way to increase market share. Instead, design teams focus their energy on particular Sales Points that will make their product outshine the competition in certain areas. Sales Points appear in column I of Figure 5-6. The values 1.5, 1.2 or 1.0 are selected (column I). The 1.5 is reserved for a demanded quality that will distinguish your product from the competition's. The demanded quality would be part of a sales campaign to promote your product or service. A 1.2 is reserved for the nice to have but not critical. Sales Points greater than 1.0 are seen as an opportunity for management to influence the design. Organizations do not have the resources to improve the satisfaction for all demanded qualities. Akao suggests a maximum of three Sales Points greater than 1.0 to emphasize the unique impact of those demanded qualities.

The DQ Composite Importance (column J) is equal to demanded quality Importance times Ratio of Improvement times Sales Points:

$$J = A \times H \times I$$
$$= 4 \times 1.25 \times 1.0 = 5.0$$

This process is repeated for all the demanded qualities. Figure 5-6 contains the traditional columns for this QFD analysis, but teams should feel free to add new ones to better meet their particular project's needs, e.g., warranty or complaints in column F.

Workshop 5-3

Take piece 3 and place it to the right of pieces 1 and 2, as shown above.

Select Target values and Sales Points. Calculate the Ratio of Improvement and the DQ Composite Importance.

Complete one column before doing the next column. Column decisions must be looked at in total. Improving every demanded quality is not rational.

Figure 5-7

DEMANDED QUALITY	A Importance	B XYZ	C Them 1	D Them 2	E Them 3	F Useful Data	G Target	H Ratio of Improvement	I Sales Points	J DQ Composite Import.	K % Composite Import.
		Customer Evaluation									
Smear-Free	4	4	5	5	5		5	1.3	1.0	5.0	6
Common Markers	19	4	5	5	3		5	1.3	1.5	35.6	43
Freestanding	2	5	2	3	1		5	1.0	1.0	2.0	2
Stays on Wall	12	3	4	2	1		5	1.7	1.5	30.0	36
Easily Removed from Wall	3	4	3	2	2		4	1.0	1.2	3.6	4
Opens Easily	1	2	3	3	5		4	2.0	1.0	2.0	2
Protects	2	2	3	2	5		5	2.5	1.0	5.0	6
Weighted Satisfaction		156	185	158	112		211				

Figure 5-8

Some Added Thoughts

Three additional analyses provide useful information.

The first is calculating the customer's weighted satisfaction with the competing products (Figure 5-8). XYZ is much better than Them 1 for Freestanding, the 5 indicating higher satisfaction than the 2. However, Them 1 has more satisfied customers for the demanded quality Stays on Wall. Them 1's 4 is larger than XYZ's 3. Unfortunately, the fact that XYZ is doing better for the unimportant Freestanding does not help market share.

XYZ Weighted Satisfaction is found by multiplying column A Importance by column B relative satisfaction.:

$$156 = 4 \times 4 + 19 \times 4 + 2 \times 5 + 12 \times 3 + 3 \times 4 + 1 \times 2 + 2 \times 2$$

The sum of these seven products is the highlighted 156 at the bottom of column B. Next, the team uses the same process for the three competitors.

Them 1 has the best product in the eyes of the customer with 185, while XYZ and Them 2 are about the same with 156 and 158, respectively. Remember the source of the numbers because orders of magnitude are needed for decision making. The new target has 261 as the forcasted satisfaction.

If all the demanded qualities are listed in the matrix, column K permits the team to say that 43% of the customer satisfaction comes from being able to use common markers. The first graph to the right in Figure 5-9 is a bar graph of the % Composite Importance from column K. The second graph on the right compares XYZ and Them 2 by plotting the satisfaction in columns B and C. This provides a visual understanding of the comparisons.

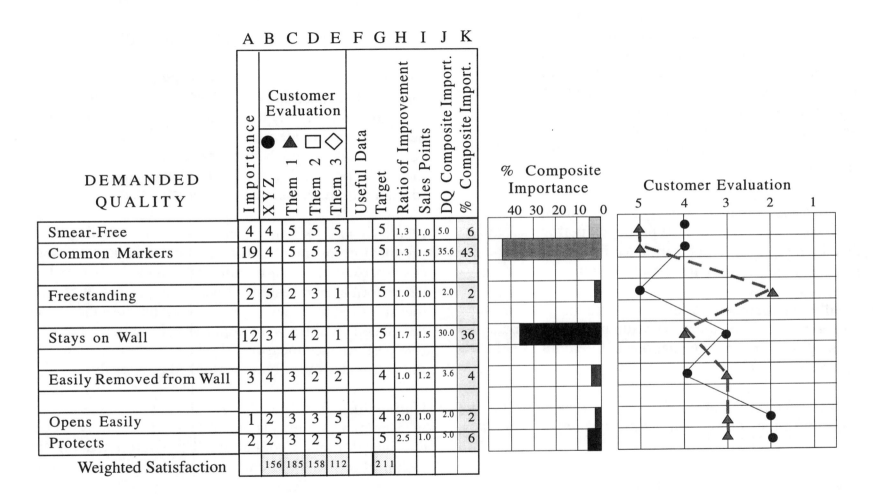

DEMANDED QUALITY	A Importance	B XYZ	C Them 1	D Them 2	E Them 3	F Useful Data	G Target	H Ratio of Improvement	I Sales Points	J DQ Composite Import.	K % Composite Import.
Smear-Free	4	4	5	5	5		5	1.3	1.0	5.0	6
Common Markers	19	4	5	5	3		5	1.3	1.5	35.6	43
Freestanding	2	5	2	3	1		5	1.0	1.0	2.0	2
Stays on Wall	12	3	4	2	1		5	1.7	1.5	30.0	36
Easily Removed from Wall	3	4	3	2	2		4	1.0	1.2	3.6	4
Opens Easily	1	2	3	3	5		4	2.0	1.0	2.0	2
Protects	2	2	3	2	5		5	2.5	1.0	5.0	6
Weighted Satisfaction		156	185	158	112		211				

Figure 5-9

85

Performance Measures

Up to this point, all information has related to the customer's demanded qualities. This information is too fuzzy for the design process because the language of the customer is not very specific. Words such as quickly, and quietly, efficiently must be changed to the engineer's language, such as time to complete or a frequency range. Measurable performance is necessary to evaluate alternative designs and to predict the satisfaction of the customer.

A performance measure is a technical measurement evaluating the product's performance of a demanded quality. A performance measure also measures the performance of a function.

Ideally, the method chosen to measure the performance can be done in the laboratory without the consumer. This allows for rapid evaluations before doing field trials of a new design.

This translation of subjective consumer statements into objective engineering terms can be challenging, but the effort is rewarding.

If you employ the traditional ways of measuring performance without using the customer's words to drive performance measurement, the chance of innovation is reduced. There is an expression:

IF YOU ALWAYS DO
 WHAT YOU DID BEFORE
YOU WILL ALWAYS GET
 WHAT YOU GOT BEFORE.

At least one performance measure should be identified for each demanded quality. There is a tendency at this point to move directly to the solutions. Self-control is necessary **only** to identify ways of measuring performance. It may become necessary to develop a means for performing an identified performance measure (design requirements, quality characteristics).

Identifying performance measures often leads directly to product improvements. For example, the manufacturer of prepackaged popcorn with 70 demanded qualities had 92 performance measures. Five of the performance measures required development work because the related demanded qualities were important. Looking at these five performance measures showed the design team important areas they had overlooked previously.

During difficult financial times, the development of measurement systems is one of the first activities to be eliminated from corporate functions. Before making any technical evaluations, some means must be found to take measurements. There are some exceptions to this, because issues such as taste may require an expert whose judgment functions as the standard measuring device.

If you cannot measure something,

Then you do not understand it!

Lord Kelvin

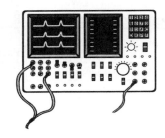

If the list of performance measures is large, the affinity diagram should be employed to reduce the set to a more manageable number by using group headings. The same level of detail as the demanded quality should be used for the columns. This is a general suggestion for all QFD matrices.

As with the consumer needs, the number of columns should not exceed 20 or 30. Today, several North American corporations are trying to interpret matrices with 100 rows and columns. Looking at the 10,000 intersections is overwhelming. Either the teams have too much detail or they should consider breaking up the data into separate matrices.

DEMANDED QUALITY	PERFORMANCE MEASURE	(Test Procedures)
Smear-Free	Performance of Pens	Write on with 12 Pens
Common Markers	Performance of Pens	Write on with 12 Pens
Freestanding	Ease of Tipping	Backboard Stability
Stays on Wall	Time on Walls	Sheet Removal
	Number of Walls	Shear Various Surfaces
Easily Removed from Wall	Effort to Remove	Peel Energy
Opens Easily	Steps to Remove	Packaging Test
Protects	Ease of Damage	Packaging Test

Notice that the demanded quality of Common Markers has the same performance measure as Smear-Free. The demanded quality Stays on Wall has two performance measures, what you measure not how.

Test procedures are defined before technical benchmarking begins (step 7 in Figure 5-1). Two examples are Write on with 12 Pens and Shear to Various Surfaces in the document for easel pad performance measures (Figure 5-11).

An extensive database identifies the critical markers used on easel pads. The test procedures describe the effort and direction for drawing lines on the paper. The standard is used to evaluate feathering and smearing.

A cause and effect diagram (fishbone) can be made for each demanded quality. The demanded quality is the effect and the performance measure is the cause. The measured values for the performance measures Sheet Removal and Shear to Various Surfaces are predictors for the customer's satisfaction with the demanded quality Stays on Wall.

Shear is used to measure the adhesive ability to stick to a variety of surfaces. Surfaces are identified by the sales force while visiting customer locations.

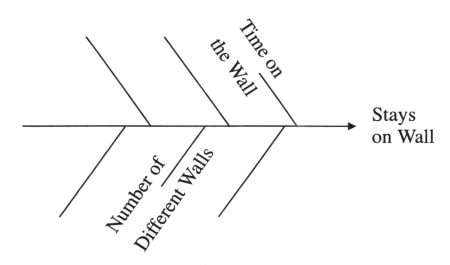

Figure 5-10

EASEL PAD PERFORMANCE MEASURES

Write on with 12 Pens

1. Test 12 various markers for feathering and smearing. Be sure to include both water- and solvent-based pens.

2. Check next sheet(s) down for bleed-through.

3. Check for opacity.

4. See if there is any difference writing on release-coated paper.

TARGET: no bleed-through, smearing, feathering.

TO DO: Is there to be absolutely no bleed-through, or is some bleed-through (with different types of markers, for example) allowed?

Shear to Various Surfaces

1. Mount surfaces on steel panels, apply sample and mount panel to shear testing stand.

A. Fabric	F. Wood paneling
B. Metal/stainless steel	G. Chalkboard
C. Cinder block (painted)	H. Whiteboard
D. Wallpaper	I. Glass
E. Paint	

2. Record time to failure.

TARGET: Unknown.

TO DO:

1. Compile available data for package book.

2. Modify tape test to use for this product.

3. Correlate this to full sheet hang test.

Figure 5-11

Workshop 5-4

Take piece 4 and place it to the right of piece 1, as shown above.

Use a separate sheet of paper with a fishbone diagram or a list for each demanded quality. Try both approaches for one demanded quality. At least one performance measure should be identified for each demanded quality in the workshop.

Record a total of 8 to 10 performance measures for this workshop.

Figure 5-12

DEMANDED QUALITY	Write on with 12 Pens	Sheet Removal	Shear Various Surfaces	Peel Energy		Backboard Stability		Packaging Test	
Smear-Free	◐	•	•	•	•	•	•	•	•
Common Markers	●	•	•	•	•	•	•	•	•
	•	•	•	•	•	•	•	•	•
Freestanding	•	•	•	•	•	●	•	•	•
	•	•	•	•	•	•	•	•	•
Stays on Wall	•	•	◐	●	●	•	•	•	•
	•	•	•	•	•	•	•	•	•
Easily Removed from Wall	•	•	●	•	•	•	•	•	•
	•	•	•	•	•	•	•	•	•
Opens Easily	•	•	•	•	•	•	•	●	•
Protects	•	•	•	•	•	•	•	○	•

Predictive quality of Performance Measure

●	strong	9
◐	medium	3
○	weak	1
•	none	0

Figure 5-13

Relationship Matrix

The predictive relationship between performance measures and demanded qualities is critical for the transformation of demanded qualities into objective design language. Identifying the strength of each performance measure's predictive ability for each demanded quality is rarely part of traditional design processes.

For each row and column intersection ask:

🗝 If I know the value for performance measure X, how well will it predict the customer's satisfaction with the product's ability to satisfy demanded quality Y?

Four options are offered:

● a strong relationship with a value of 9.

◐ a medium relationship with a value of 3.

○ a weak relationship with a value of 1.

• no relationship with a value of 0.

The filled circle, the quarter-filled circle and the empty circle are used instead of weights. These symbols facilitate identifying patterns of relationships in the matrix. In the *Consumer Reports* annual automobile issue, it is easy to see which automobiles are most desirable by simply looking at the patterns of the symbols in evaluation tables.

The small dot is a useful cell entry for large projects when there is time between discussions. It functions as a bookmark, letting the team know what was previously analyzed. The dot identifies cells with no relationships.

The same symbols are used for positive or negative relationships. A positive relationship produces a more satisfied customer when the performance value improves. A negative relationship produces a less satisfied customer when the performance value improves.

Significant discussions are recorded and flagged at the appropriate cell. QFD software allows large records to be attached to each cell. For instance, the bent page corner symbol from Excel indicates that there is further information behind the cell.

The easel pad team felt that the performance measure Write on with 12 Pens had a medium relationship with Smear-Free but a strong relationship with Common Markers.

⌐ Important demanded qualities must have a performance measure with at least a medium relationship. A strong relationship is preferred.

The gaps between 9, 3 and 1 emphasize the more important performance measures, as in the Pareto Principle (most of the value will come from the critical few). More than 50% of the cells should represent no relationship. Keeping the Pareto Principle in mind, not everything is related. Designing to the critical 20% will satisfy 80% of the customer's desires. Do not use the weak relationship for a possible relationship between a performance measure and a demanded quality. Use a question mark if you do not know whether there is a relationship. After an investigation, the appropriate symbol replaces the question mark.

If traditional measures of performance are placed in the columns instead of using the recommended process, there are usually blank rows and blank columns. A blank row means the demanded quality will not influence the design. This is serious for the important demanded qualities. A blank column indicate wasted resources measuring something that does not address customer needs. Of course there are issues that the customer is not aware of, such as governmental or corporate requirements.

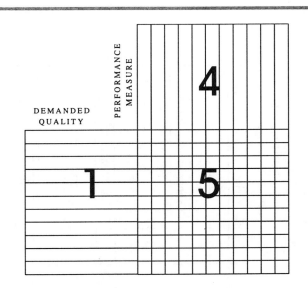

Workshop 5-5

Rotate piece 4 and place piece 5 as shown above.

For each intersection of row and column answer the question, "If I tell you the performance measure results, how much of an influence does it have on the satisfaction of the customer?"

Discuss the relationship between your demanded qualities and the performance measures. Enter the appropriate symbol.

Figure 5-14

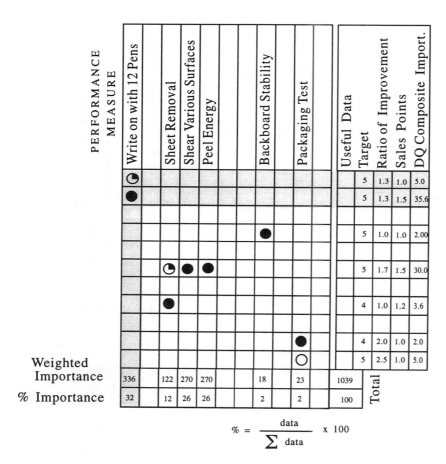

Figure 5-15

$$\% = \frac{data}{\Sigma\ data}\ x\ 100$$

Product Planning Table (prework)

Before establishing the design targets, the relative importance of the performance measures must be calculated. The results from the relationship matrix provide the link.

A weighted importance for each performance measure is found by calculating the sum of the products of the strengths of the relationships in each cell and the associated composite weight for all the rows.

The Weighted Importance score for the Performance Measure in the first column is equal to the sum of the product of the strength of the relationship in the first column times the related DQ Composite Importance score in the last column:

Weighted Importance = ◔ x 5.0 + ● x 35.6

= 3 x 5.0 + 9 x 35.6 = 335.4

Only two rows in the first column have a nonzero value because there is some predictive relationship. The 335.4 was rounded up to 336, for ease of reading.

This process is used for all columns.

Calculating the % Importance is more understandable than Weighted Importance. To find the % Importance calculate the total of the Weighted Importance row. The rounded 1039 is shown at the right end of the row.

The % Importance for the first column is the raw score 336 divided by 1039 multiplied by 100:

Write on with 12 Pens % Importance =

$$(336 / 1039) \times 100 = 32$$

It is clear that the 32 for Write on with 12 Pens is more important than the 12 for Sheet Removal, but not very different than the 26 for Peel Energy. More precision than rough ranking is neither necessary nor realistic.

Some organizations plot a histogram or bar chart below the data for quicker interpretation of the results.

If these few performance measures represented all the dimensions of quality related to repositionable easel pad sheets, then 32% of the design contribution toward satisfying the customer comes from how well 12 different markers left their mark.

The 32% is the largest Importance score. A team may want to stop at this point and concentrate design efforts on improving the performance of Writing on with 12 Pens. This could be a serious mistake.

The current design may be performing well for the Write on with 12 Pens when compared to the competition. The technical benchmarking on Peel Energy may indicate a large deficiency when compared to the competition. These issues will be covered after finding the % Importance for your case.

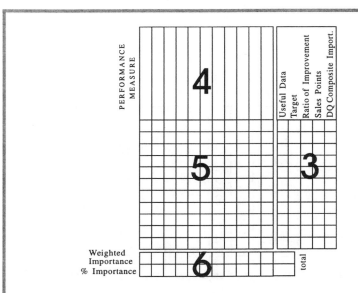

Workshop 5-6

Place pieces 4, 5, 6 and 3 as shown above.

Calculate the Weighted Importance and the % Importance for each of the Performance Measures.

Find the total. Convert the raw numbers to % by dividing by the total and multiplying by 100. This is also called the normalized score.

Figure 5-16

PERFORMANCE MEASURE	Write on with 12 Pens		Sheet Removal	Shear Various Surfaces	Peel Energy			Backboard Stability		Packaging Test		Total
Weighted Importance	336		122	270	270			18		23		1039
% Importance	32		12	26	26			2		2		100
Measured Performance: XYZ	5		0	75	215			2		P		
Measured Performance: Them 1	0		2	110	181			3		P		
Measured Performance: Them 2	0		2	108	76			6		F		
Measured Performance: Them 3	7		0	62	40			0		P		
	# Failures		Newtons	Hours	Ergs			Newtons		Pass/Fail		

Figure 5-17

Product Planning Table

The Product Planning Table sets targets for the design requirements and prioritizes the development efforts.

Technical benchmarking uses the performance measures to evaluate the performance of your organization's current design against the competition. These should be measurements that do not use the consumer because you have already used the consumer for the Quality Planning Table.

The last four rows in Figure 5-17 are the results of actually conducting specific tests, developed for performance measures.

The units of measure for the Performance Measures are: number of failures, Newtons to failure, time to fail in hours, ergs, Newtons to tip and pass/fail. Avoid using a pass/fail test because this makes it more difficult to select the best direction for improvement. Remember, the measures and data in this manual were modified by the project consultants; they do not represent reality.

Coding the data with a five-point scale for quick visual understanding enhances reporting and discussion. The coding allows a plot of all performance measures with a single vertical axis. This makes relative position clearer and is similar to plotting the consumer's judgment of the competing products.

Use a 5 for outstanding performance and a 1 for poor performance. This transformation is easy for performance measures that are minimized or maximized. This transformation is difficult for the *target-is-best* performance measure if satisfaction is not symmetric for deviations from the target.

Genichi Taguchi suggests that the quality of a product depends upon average performance and variance. A single metric, called the loss function, is used to measure quality. This measure is suggested as an improvement over the average shown in this example. It is most productive to have the average, the variance and the loss function.

Some of Taguchi's approaches to design are presented in the next chapter. The material in Chapter 6 can be skipped, especially if your team is just upgrading a good design. However, this material is further developed for manufacturing in Chapter 10.

Rows can be added to the Quality Planning Table to accommodate this additional information.

We see that all the pieces form a figure that looks like a house (Figure 5-18). For this reason, this particular phase of QFD is often called the House of Quality. Some publications unfortunately call any matrix used in the QFD process a House of Quality. To minimize confusion, this collection of tables and a matrix is called the Demanded Quality vs. Performance Measure Matrix.

The roof (piece 8) is discussed on page 100.

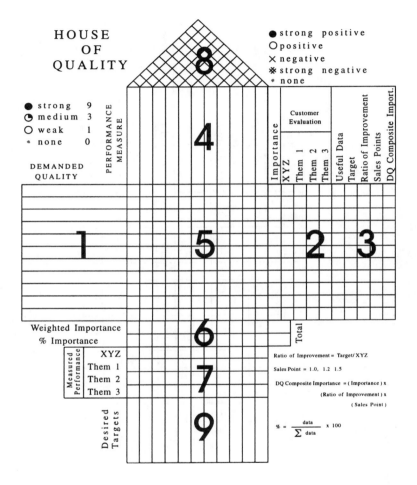

Figure 5-18

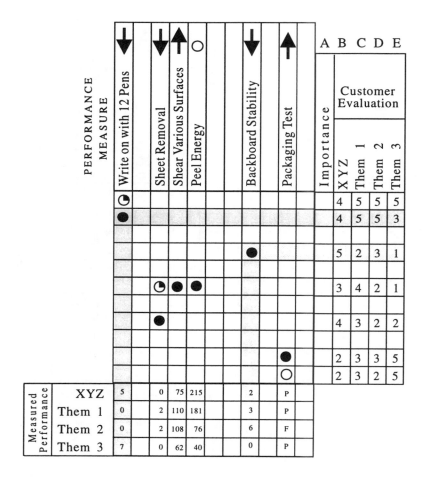

Figure 5-19

Validate Strength of Relationship

The methods for measuring the performance should be validated before going any further.

Figure 5-19 illustrates this crucial step. Any cell which has a strong relationship between performance measure and demanded quality should have agreement between the consumer's evaluation of the competing products and your measurement of the competing products.

Ford Motor Company found that adding arrows at the top of the chart to indicate the desired direction for better performance clarified discussion for comparing products and selecting targets. A circle was used for target-is-best. If a column has positive and negative relationships, respectively, with equally important demanded qualities, then target-is-best would be used for the performance measure.

The intersection of column one and row two has a strong relationship. Lower scores for column one are more desirable. The rank ordering would be Them 1 and Them 2 are best, with XYZ coming in third and Them 3 coming in fourth. Looking at the customer evaluation for row two, there is agreement with the performance measure. Thus the measure is useful for predicting the consumer's satisfaction.

This is not the case for Backboard Stability in row four. If smaller results for the performance measure are better, the ranking of the column for the competing products does not match the consumer's ranking. The method chosen to measure this performance may not be addressing the real demanded quality of the consumer.

For example, a demanded quality for a sports car might be Dry Seat. The design team selected Grams of Water on the seat as the performance measure generating a strong predictive value. Consumer feedback reported the current performance as horrible, while the team's tests indicated that it was the best in its class. How could this be?

The test engineer was considering the moisture that may get in around the door when a tractor trailer produces a wave of water over the car. The test procedure measured the water on the seat after spraying a fire hose at the closed doors and windows.

The customers delivered a poor subjective rating because the car's jellybean styling makes the seats wider than the roof. When it rains and the door is opened, the seats get wet because the exterior rain gutter no longer exists.

In this case finding a better test plan was easy, but sometimes a substitute quality characteristic must be found. The most common ones relate to accelerated aging for reliability testing. For instance, putting a soft plastic in an oven is supposed to simulate time for the migration of plasticizors.

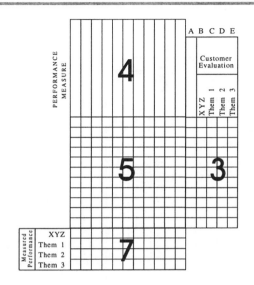

Workshop 5-7

Place pieces 4, 5, 7 and 3 as shown above.

Use arrow and circle to record the direction of improvement at the top of piece 4.

Since you have not measured data for the performance measures, it is not possible to check the validity of the strong relationships.

If you have measurements, then search for the cause of any inconsistency.

Figure 5-20

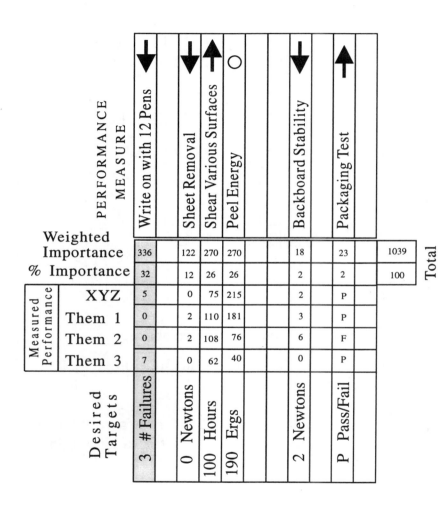

Figure 5-21

Selecting Target Values

Projects designing a model upgrade for an existing systems design should investigate design conflicts among the performance measures. This is covered in the next section. These conflicts would force tradeoffs between competing target values. The process is unique to each organization, but the following must be considered:

1. How important is a performance measure?
2. How does the organization compare to the competition?
3. How does the performance measure relate to the corporate image?
4. What are the organization's technical abilities?
5. What resources are available?
6. What do you think the competition is developing?

Sometimes the most important performance measure may not get a target for improvement because the product is already better than the competition and there do not seem to be any new developments in the near future.

After selecting the target values, add an additional row to identify the priority of the projects (Figure 5-22). If the competition is way ahead on some measure, buying the technology may be the best decision. Your resources could then be more productively focused on your strengths.

98

The team decided on a two-step improvement for the most important performance measure. First, reduce the number of failures of the Write on with 12 Pens to 3 for the next model and then go for zero for the later model.

If Peel Energy is a predictor for Stays on Wall, what is the best target value (Figure 5-13)? In Figure 5-8, we see that Stays on Wall is considered best for Them 1's product by the consumer, but not world class. Looking at the completed House of Quality in Figure 5-25 provides all the information that previously required looking at three tables. In Figure 5-22, it appears that 215 for XYZ is too high and, possibly, 181 for Them 1 is too low. It was felt that 190 ergs was a more desirable target.

To help decide which project should be initiated, four additional rows can be added.

The first row is simply a ranking of the % Importance.

The second row is a five-point scale of technical difficulty to attain the target value. A 5 is very difficult and a 1 is available internally or externally.

The third row represents current manufacturing capability for the current measured performance. The Cp index (ratio of specification to manufacturing 6σ) or the Cpk index (the number of process σ to the nearest specification / 3) can be used.

A check mark in the last row indicates a committed projects. An X indicates that no special effort is planned.

	# Failures	0 Newtons	100 Hours	190 Ergs		2 Newtons	P Pass/Fail		Total
Weighted Importance	336	122	270	270		18	23		1039
% Importance	32	12	26	26		2	2		100
XYZ	5	0	75	215		2	P		
Them 1	0	2	110	181		3	P		
Them 2	0	2	108	76		6	F		
Them 3	7	0	62	40		0	P		
Desired Targets	3	0	100	190		2	P		
Rank	1	3	2	2		4	4		
Difficulty	3	1	5	1		2	1		
Current Cp	1.2	2.0	1.0	3.6		0.9			
Selected	✓	✓	X	✓		X	X		

(Measured Performance rows: XYZ, Them 1, Them 2, Them 3. Desired Targets row units: 3 # Failures, 0 Newtons, 100 Hours, 190 Ergs, 2 Newtons, P Pass/Fail)

Figure 5-22

Identifying Performance Measure Conflicts

Model upgrade projects start with an existing design system. Performance measures in existing designs often conflict with each other. The peak in the House of Quality (Figure 5-23) documents positive or negative influences between the performance measures. If there is a negative or strongly negative impact between performance measures, the design must be compromised UNLESS THE NEGATIVE IMPACT IS DESIGNED OUT.

Classical TRIZ (Theory of Inventive Problem Solving) was built upon the observation that only 40 principles were used to resolve many of the contradictions solved in world patents. Compromises in the target values are not necessary. The use of TRIZ offers an exciting approach to systematic innovation and is presented in Chapter 8.

Some conflicts cannot be resolved because they are issues of physics. For example, more power and smaller size are in conflict. Others are design-related, leaving the team a degree of choice about how to resolve them. Adding a roof to the Product Planning Table will allow this analysis and link it to the selection of target values.

The cells in the roof document the influence between performance measures. Five symbols are used, with the following meaning:

● strong positive
○ positive
✕ negative
✱ strong negative
• none

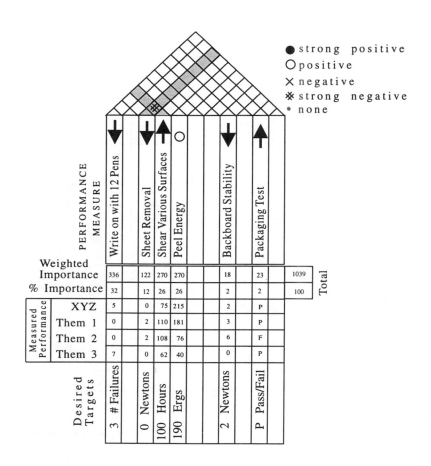

Figure 5-23

The negative can also represent constraints.

Constraints may not be bidirectional. As a result, improving one of them may negatively impact another, but in the other direction there may be no impact. The cells are split to record this information. For example in designing a pencil, the smaller grip effort would negatively impact a dark line, but darker lines would have no effect on grip force. The cell looks like:

Asking the following question helps to clarify relationships among design measures:

"If performance measure X is improved, will it help or hinder performance measure Z?"

Note that this is not a correlation. Consider two cases:

1. Both larger values for X and larger values for Z are better. If X is increased and causes Z to increase, then X has a positive or strong positive influence.

2. Larger values for X but smaller values for Z are better. If X is increased and causes Z to decrease, then X has a positive or strong positive influence.

Figure 5-23 shows a strong negative relationship between Sheet Removal and Shear Various Surfaces.

The performance measure Sheet Removal is not very important, but care must be taken in the selection of target values. The target value selected may make it impossible for a desired Shear to Various Surfaces to be attained. Compromising on Sheet Removal may be necessary.

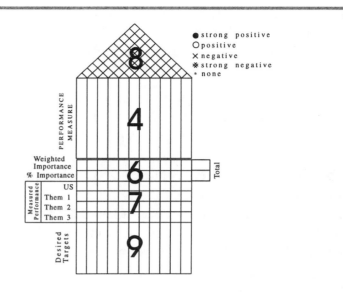

Workshop 5-8

Place pieces 9, 7, 6, 4 and 8 as shown above.

Discuss the influence between all pairs of performance measures. Ask the question:

"If Performance Measure X is improved, will it help or hinder Performance Measure Z?"

Use the five symbols indicating the influence of one performance measure upon another. If the influence is not bidirectional, enter two symbols. Place the symbol on the left for the left PM improving first. If the right PM is improved first, place the symbol to the right.

Figure 5-24

For a large project, the roof will at least raise a flag and identify two design units that must work together; otherwise there will be a design conflict.

 The output from the analysis of the House of Quality is the selection of the few critical new and important performance measures.

The House of Quality (Figure 5-25) functions as a living document and a source of ready reference for related product and future model upgrades. Once you become comfortable with the process, it is more efficient to use a spread sheet or QFD software that represents the complete House of Quality. Many companies are satisfied that this information is all they need from QFD. At this point they have identified the critical performance measures and associated target values. They have reached the heart of their journey toward designing a world-class product.

Some organizations find this is all they need from QFD and return to their traditional design process.

It is time to

and review your understanding.

Multiple Customers

Most QFD projects have multiple customer segments. Three methods of working with different segments are offered for different applications. The prioritization of the segments facilitates selecting the appropriate subset of all customer segments.

There are two types of customer segments. One group includes those involved in the time line of decisions to buy or use a product. A single product must be designed to satisfy all their needs.

The other group includes the different users in their various environments. There are three options for satisfying the different segments: make only one product for all customer segments, make one product with several options or adjustments or develop a product line.

Each option requires gathering the demanded qualities for all the selected customer segments. The following three options offer different ways to integrate this information into the design process.

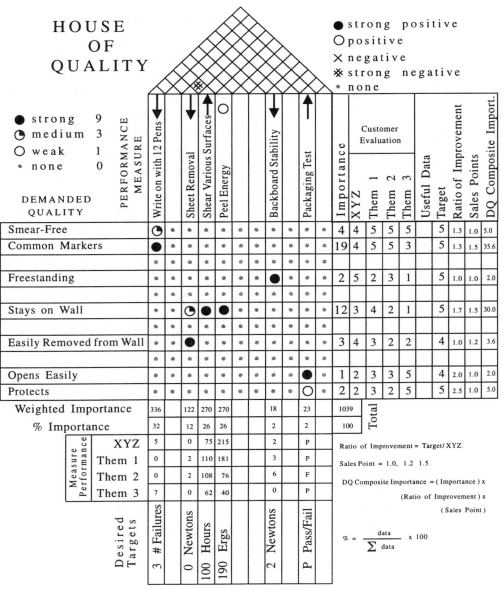

Figure 5-25

DEMANDED QUALITY	Segment 1				Segment 2				Segment 3			
	Importance	XYZ	Them 1	Them 2	Importance	XYZ	Them 1	Them 2	Importance	XYZ	Them 1	Them 2

Figure 5-26

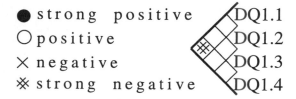

● strong positive　DQ1.1
○ positive　DQ1.2
✕ negative　DQ1.3
※ strong negative　DQ1.4

Figure 5-27

If the demanded qualities are identical, only one matrix for translating the demanded quality into performance measures is needed. The right-hand side of the Importance and Customer Evaluation columns would be repeated for each segment (Figure 5-26). Since the customer segments have diverse priorities and evaluations, different rankings of performance measures would be shown at the bottom of the matrix. Then the team makes a judgment call for each of the target values.

The decision to make one product with options or different products depends on the number of conflicts among customer segments. Before checking for conflicts among the segments, a check for conflicts between the demanded qualities for each segment is necessary. The roof in Figure 5-25 can be rotated 90 degrees and used for this testing. A conflict is shown between Demanded Quality 1 (DQ1.1) and Demanded Quality 4 (DQ1.4) for Segment 1 (Figure 5-27).

Combining the checking for such conflicts between two segments forms a square. The square in Figure 5-28 is used to check for conflicts between segments; note the very strong conflict between the Demanded Quality 4 from Segment 1 (DQ1.4) and Segment 2 (DQ2.4). The team decides whether these conflicts require different products or whether they can be satisfied by an adaptable single product.

In Chapter 3, AHP was used to calculate the customer segment priorities. The priorities can now be used for decision making.

Working on a three-dimensional matrix can help this decision. The vertical axis has the Demanded Qualities (DQ). The left-directed axis has the customer segments (SEG). The left face records the importance of each DQ for each SEG. This example uses a five-point scale. A 5 means the DQ has great importance and a 1 means it has little importance.

The right-directed axis contains current design concepts (CON). The right face contains the ability of each CON to satisfy each DQ. A 5 now means the design is very effective in satisfying the DQ, and a 1 means the design does a poor job of satisfying the DQ.

The top face contains the overall performance of each concept for each segment. The 32 at the intersection of SEG3 and CON1 is the weighted performance for the concept. The weighted performance is the sum of the products of the performance for each demanded quality and the importance for each demanded quality:

$$32 = 1x2 + 4x3 + 5x1 + 2x4 + 1x5$$

From the top face, a product mix of CON1 for SEG 2 and CON2 for SEG 3 is most effective. To find one product's overall satisfaction for each concept, find the sum of its scores for each concept. The 82 shown for CON2 is better than the 67 for CON1. This calculation assumes that the segments are equally important. If the segments are not equally important, a weighted sum is calculated. Each weighted performance is multiplied by the segment importance before calculating the sum.

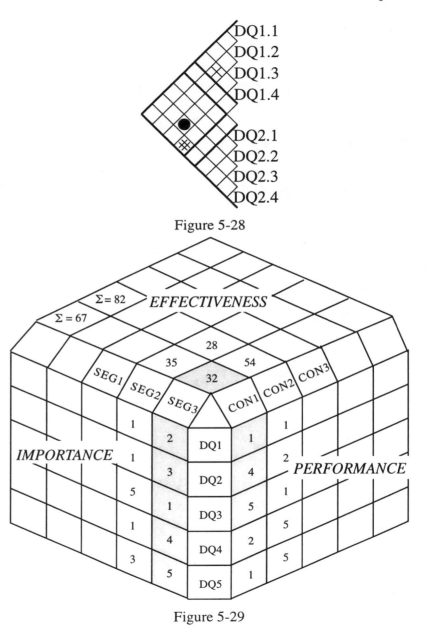

Figure 5-28

Figure 5-29

For organizations that continue the QFD process, the next chapter explores Taguchi's design philosophy. This will focus on his loss function, a comprehensive way of measuring quality, thereby further ensuring customer satisfaction. If your team is designing an upgrade, you may choose to skip ahead to Chapter 7, which explores alternative design selection.

Keep on Stepping.

Chapter 6

A Better Way to Measure Quality

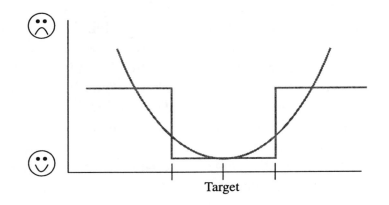

Target

Upon completing this chapter, you will understand:

♦ The philosophy behind Robust Design.

♦ How to reduce the impact of uncontrolled influences upon the performance of your product.

Introduction

Dr. Genichi Taguchi's impact upon North American product design and manufacturing processes began in November 1981. Ford Motor Company requested that Dr. Taguchi make a presentation. I was fortunate to be invited to hear about this powerful design technique. A different method of measuring quality is central to Taguchi's approach to design. **Loss function** measures quality. The loss function establishes a financial measure of the user dissatisfaction with a product's performance as it deviates from a target value. Thus, both average performance and variation are critical measures of quality. Selecting a product design or a manufacturing process that is insensitive to uncontrolled sources of variation improves quality. Dr. Taguchi calls these uncontrolled sources of variation **noise factors**. This term comes from early applications of his methods in the communications industry. Applying Taguchi's concept entails evaluating both the variance and the average for the technical benchmarking in QFD. The loss function provides a single metric for comparison.

Static Taguchi applications search for a product design or manufacturing process that attains **one** fixed performance level. A static application for an injection molding machine finds the best operating conditions for a single mold design. Dynamic applications use mold dimensions as the signal and search for operating conditions which yield the same percentage shrinkage for any dimension in any orientation. The dynamic approach allows an organization to produce a design that satisfies today's requirements but can be easily changed to satisfy tomorrow's demands. You can consider this latter approach as contingency planning for some unknown future requirement. In dynamic applications, a **signal factor** moves the performance to some value and an **adjustment factor** modifies the design's sensitivity to this factor. If you plot a straight line relationship, with the horizontal axis as the signal factor and the vertical axis as the response, the adjustment factor changes the slope of the line. Being able to reduce a product's sensitivity to changes in the signal is useful. For example, if you are designing a sports car, your desired outcome might be a car that allows the driver to change the feel of the road. The signal factor would be a control knob setting. The analysis could determine that the suspension system is the adjustment factor. The adjustment factor adjusts the magnitude of change in road feel to a given change in the knob setting. Several other design specifications would assure a predictable relationship in the control knob setting and the feel. Changes in road conditions and weather would have minimal effect upon the relationship between knob adjustment and feel of the road.

Listening to the voice of the customer helps organizations create good systems designs. Some teams use focus groups to gather input for these designs. *Fortune Magazine* (April 1995) printed an article by Justin Martin entitled "Ignore Your Customer." It presents several examples of products that were strongly not purchased by consumers. The author suggests that studying the customers under **natural** conditions would provide additional useful information. Going to the *Gemba* is a crucial step in QFD. This has been advocated since 1985. The *Gemba* is the total environment in which the customer lives and works.

We saw the Voice of the Customer Table with three components: customer verbatim response, context of use and integration of verbatim response and context (Chapter 4, Figure 4-5). Eventually, the team divides this expanded list of customer information into demanded qualities, failure modes, solutions, etc. The Context of Application Table (Figure 4-3) identifies some of the environmental sources of uncontrolled variation in product performance. Some examples of sources of variation for an easel pad are the force applied to the paper, humidity and whether is it being used inside or outside. Taguchi's Robust Design reduces the impact of uncontrolled sources of variation upon the product's performance.

How to Measure Quality

Traditionally, quality is viewed as a step function, as shown by the heavy line graph in Figure 6-1. A product is either good or bad. This view assumes a product is uniformly good between the specifications (LS is the lower specification and US the upper specification). The left vertical axis represents the degree of displeasure the customer has with the product's performance. Curves A and B represent the frequencies of performance of two

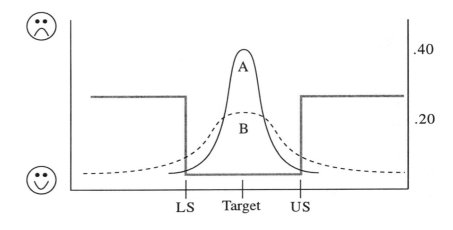

Figure 6-1

designs during a certain time period. B has a higher fraction of "bad" performance and therefore is less desirable than A. The ordinate is shown in the right vertical axis, but will not be shown in the later figures.

Sometimes traditional decision makers and those using Taguchi's loss function will make the same judgments. If organizations consider both the position of the average and the variance, and if the averages are equal and/or the variances are equal, then the traditional decision maker and one using Taguchi's loss function will make the same decision. However, the traditional decision maker calculates the percent defective over time when both the average and variance are different.

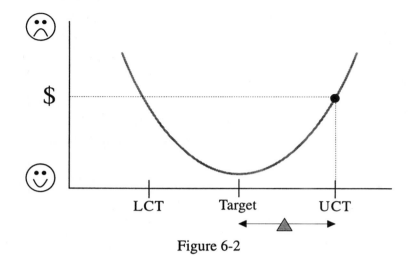

Figure 6-2

Taguchi believes that the customer becomes **increasingly** dissatisfied as performance departs farther away from the target.

He suggests a quadratic curve to represent a customer's dissatisfaction with a product's performance. The quadratic curve is the first term when the first derivative of a Taylor Series expansion about the target is set equal to zero. The curve is centered on the target value, which provides the best performance in the eyes of the customer. Identifying the best value is not an easy task. Targets are sometimes the designer's best guess.

LCT represents lower consumer tolerance and UCT represents upper consumer tolerance. This is a customer-driven design rather than an engineer's specification. Experts often define the consumer tolerance as the performance level where 50% of the consumers are dissatisfied. Your organization's particular circumstance will shape how you define consumer tolerance for a product.

The equation for the target-is-best loss function uses both the average and the variance for selecting the best design. The equation for average loss is:

$$\overline{Loss} = k\{\sigma^2 + (\overline{y} - T)^2\} \quad \text{where} \quad k = \frac{\$}{\Delta^2}, \text{T = target}$$

σ^2 = performance variance and $\overline{y}$ = performance average

Calculating the average loss permits a design team to consider the cost-benefit analysis of alternate designs with different costs yielding different average losses. As seen in Figure 6-2, there is some financial loss incurred at the upper consumer tolerance. This could be a warranty charge to the organization or a repair expense.

Most applications of the loss function in QFD can use a value of 1 for k since the constant would be the same for all competitors as it relates to the customer.

The graphics show a symmetric loss about the target, but this is not always the case.

If two products have the same variance but different averages, then the product with the average that is closer to the target (A) has better quality (Figure 6-3).

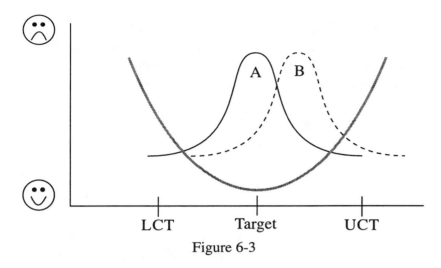

Figure 6-3

If two products have the same average but different variance, then the product with the smaller variance has better quality (Figure 6-4). Product B performs near target less often than its competitor.

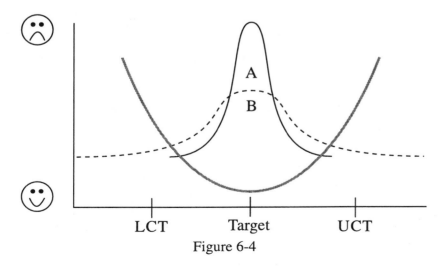

Figure 6-4

What if both average and variance are different? Calculating the average loss assumes you agree with the concept of the loss function. The product with smaller loss has the better quality (Figure 6-5). If curve A is far to the right, then curve B would be the better. If curve A is centered on the target, then curve A would be better. Somewhere in between, both have the same loss.

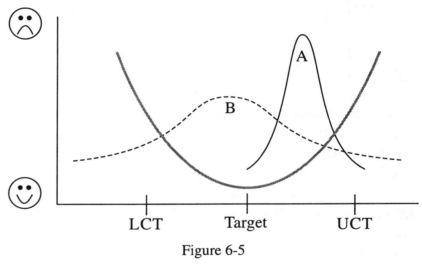

Figure 6-5

Loss Function and Technical Benchmarking

Teams should gather data collected for technical benchmarking in a real environment. A real environment is one in which everything is not controlled and ideal. Our product and the competitor's product would be evaluated at different temperatures, humidities and other conditions. The laboratory can simulate these conditions. By evaluating the product's performance in several environmental conditions, you would have realistic data to calculate the real-world variance.

An orthogonal array can define a balanced study of different environmental conditions. The two or three important environmental conditions, each at two levels, provide a good estimate of the environmental variation. The humidity is represented by H, the weight of items taped to the sheet on the wall by W and the surface texture by T.

Environment	H	W	T
1	1	1	1
2	1	2	2
3	2	1	2
4	2	2	1

The 1 and 2 under H represent high and low humidity. The four different combinations of environments are used to determine the average and variance of each product's performance. Instead of using all eight different combinations, the orthogonal array uses a special subset of the eight. Due to the balanced nature of these four combinations, the effect of the missing four can be predicted.

H	W	T
1	1	1
1	1	2
1	2	1
1	2	2
2	1	1
2	1	2
2	2	1
2	2	2

Another option is to select the best and the worst environmental combination of the eight combinations. This approach further reduces the number of environments evaluated to two.

The average loss for the data is:

$$\bar{L} = k \frac{1}{n} \sum_{i=1}^{n} (y_i - T)^2 \quad \text{where} \quad k = \frac{\$}{\Delta^2}$$

The calculations of the variance and loss are entered in two additional rows of the matrix (Figure 6-6). The ratio of the average loss of one competitor to another is independent of k. The information of the average, variance and loss ratio identifies the directions for improvement as defined by the average loss equation.

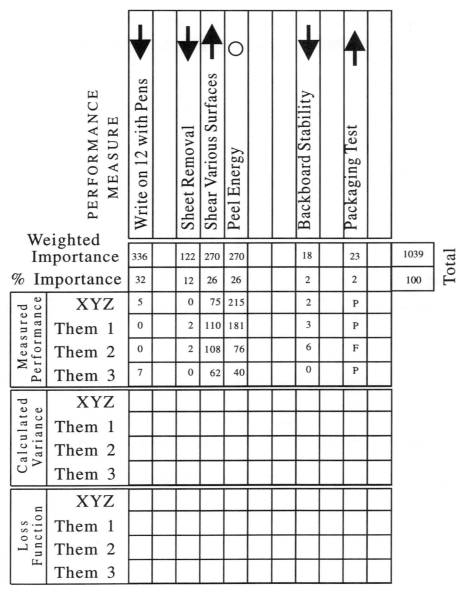

	PERFORMANCE MEASURE										
		Write on 12 with Pens ↓		Sheet Removal ↓	Shear Various Surfaces ↑	Peel Energy ○			Backboard Stability ↓	Packaging Test ↑		Total
Weighted Importance		336		122	270	270			18	23		1039
% Importance		32		12	26	26			2	2		100
Measured Performance	XYZ	5		0	75	215			2	P	
	Them 1	0		2	110	181			3	P	
	Them 2	0		2	108	76			6	F	
	Them 3	7		0	62	40			0	P	
Calculated Variance	XYZ										
	Them 1										
	Them 2										
	Them 3										
Loss Function	XYZ										
	Them 1										
	Them 2										
	Them 3										

Figure 6-6

The loss function curves for larger-is-better and smaller-is-better have a similar message when comparing different designs which have different averages and variances.

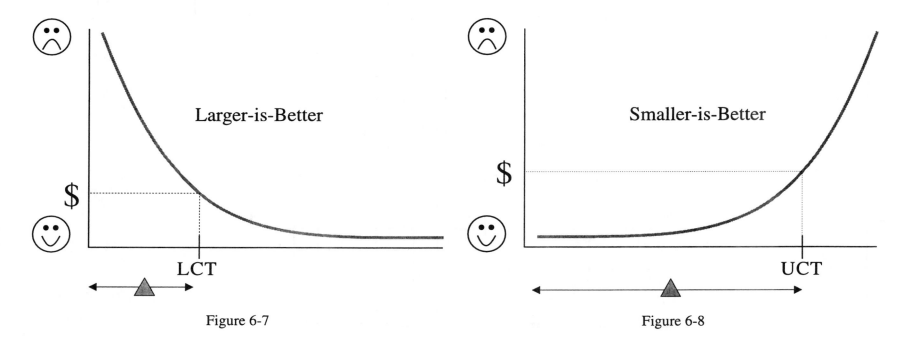

Figure 6-7   Figure 6-8

The calculation for the average loss for the data collected for the larger-is-better type performance measure is:

$$\overline{L} = k \frac{1}{n} \sum_{i=1}^{n} \left(\frac{1}{y_i} \right)^2 \quad \text{where} \quad k = \$\Delta^2$$

The calculation for the average loss for the data collected for the smaller-is-better type performance measure is:

$$\overline{L} = k \frac{1}{n} \sum_{i=1}^{n} y_i^2 \quad \text{where} \quad k = \frac{\$}{\Delta^2}$$

Sometimes a team may have trouble getting their design to perform at the target value. In another scenario, a team may not be able to sufficiently reduce the cost of making the product. The team's design may also experience trouble with variance. Depending on the circumstances, the team may benefit from considering alternative design and may even choose to create new designs. This can be an exciting, creative process. The next chapter explores ways to approach this fascinating stage of product design.

It is time to

and review your understanding.

The analysis of the customer's demanded quality has been refined by the the use of the loss function. This refinement involves the generation and evaluation of alternative design concepts; this is the topic of the next chapter.

You are another step closer to your world-class product!

Chapter 7

New Design Concepts

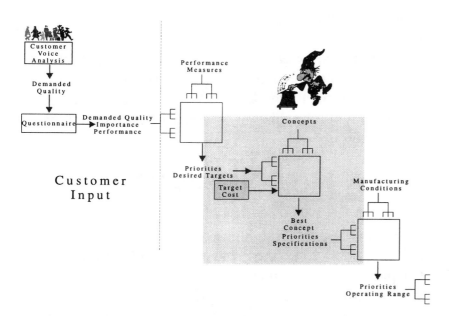

Step-by-Step QFD
for
Product Design

Customer
Input

Upon completing this chapter, you will be able to:

♦ Evaluate alternative designs.

♦ Generate new concepts.

♦ Add to your patent list.

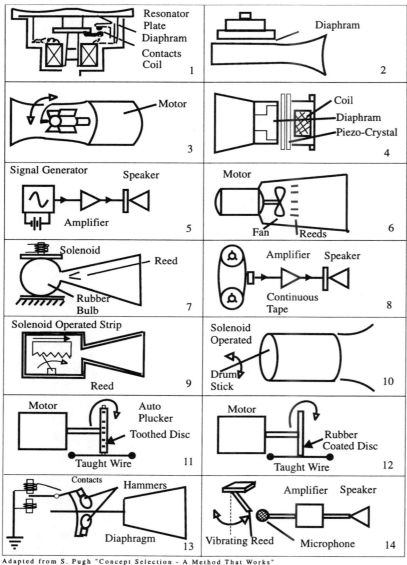

Resonator Plate / Diaphram / Contacts / Coil — 1

Diaphram — 2

Motor — 3

Coil / Diaphram / Piezo-Crystal — 4

Signal Generator / Speaker / Amplifier — 5

Motor / Fan / Reeds — 6

Solenoid / Reed / Rubber Bulb — 7

Amplifier / Speaker / Continuous Tape — 8

Solenoid Operated Strip / Reed — 9

Solenoid Operated / Drum Stick — 10

Motor / Auto Plucker / Toothed Disc / Taught Wire — 11

Motor / Rubber Coated Disc / Taught Wire — 12

Contacts / Hammers / Diaphragm — 13

Amplifier / Speaker / Vibrating Reed / Microphone — 14

Adapted from S. Pugh "Concept Selection - A Method That Works"
International Conference on Engineering Design, 1981

Figure 7-1

Pugh's Selection and the Design Process

Stuart Pugh, former Babcock Professor of Engineering Design and Head of the Design Division of the University of Strathclyde, Glasgow, developed an effective approach to the design process. Pugh's method reduces the possibility of selecting the wrong concept and increases the likelihood of selecting the best dependable design.

Pugh's Selection enables the design team to systematically evaluate different features of various designs, including cost. The value of a product to the customer is a combination of its quality, delivery and price. A simplistic model indicating the direction of improvement for achieving a higher level of customer satisfaction is:

$$Value = \frac{Quality \times Delivery}{Price}$$

Both QFD and the design process concentrate on building quality into the product. As suggested in Chapter 4, price should be driven by market considerations. The primary issue for the design team is creating a design with a cost that produces the desired profit:

$$Cost = Price - Profit$$

Pugh originally presented his selection process to the Club of Rome in 1981. It has taken many years to hear about and apply his lessons for product design. Organizations interested in continuous improvement can not afford delays in knowledge transfer. Poor communication within and among organizations inhibits the spread of useful knowledge. I witnessed an extreme example of this slow transfer of crucial information in a manufacturing plant. It took 18 months for a solution for determining the best manufacturing conditions to travel from one team to another—a distance of only 50 meters! Sharing tips and techniques designed to eliminate unproductive behavior is central to step-by-step QFD.

The first step in Pugh's Selection involves brainstorming a wide variety of alternate designs. The Theory of Innovation Problem Solving (TRIZ) would offer many alternative technologies. It is important to use some type of brainstorming process, since even a bizarre response may have a usable component. For example, during the mid 1980s, a paper honeycomb design was used to fill the hollow interior of helicopter rotors. This seemingly outlandish idea provided an innovative solution to the rotor's need for lightweight structural support. This chapter uses Pugh's original example, a car horn, to illustrate the method of evaluating and generating new concepts. Figure 7-1 shows 14 alternative designs for a car horn. You will be less restricted in your concepts if you do not consider the criteria for measurement. Alternative 10—an automated drumming of a drum—seems a little questionable. Nevertheless, there might be some small aspect of alternative 10 that can be incorporated into another design. There is a detailed information sheet for each concept.

Workshop 7-1

As a fun exercise for generating concepts, use the design of a shampoo container. This container will be used in a hotel or motel.

Who are some of the customers?

Brainstorm design ideas for containers with the hotel guest in mind.

Think of the containers you have liked and those that should not exist.

(My least favorite is a small tear packet with no tear mark.)

Figure 7-2

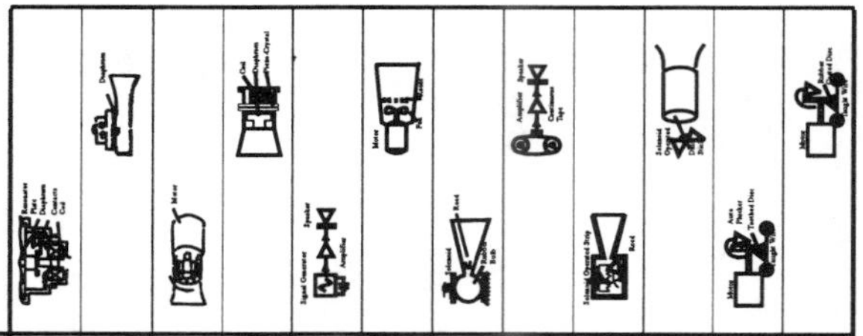

CRITERIA \ CONCEPT	1	2	3	4	5	6	7	8	9	10	11	12
106-125 dB		S	-		+	-	+	+	-	-	-	-
2000 - 3000 Hz		S	S		+	S	S	+	S	-	-	-
NO CORROSION		-	-		S	-	-	S	-	+	-	-
SHOCK RESIST	R	S	-		S	-	S	-	-	S	-	
TEMP RESIST	e	S	-		S	-	-	-	S	S	-	
RESPONSE TIME	f	S	-		+	-	-	-	-	S	-	
COMPLEXITY	e	-	+		S	+	+	-	-	-	+	+
POWER NEED	r	-	-		+	-	-	+	-	-	-	-
EASY to MAINTAIN	e	S	+		+	+	+	-	-	S	+	+
WEIGHT	n	-	-		+	-	-	-	S	-	-	-
SIZE	c	-	-		S	-	-	-	-	-	-	-
NUMBER of PARTS	e	S	-		+	S	S	-	-	+	-	-
EXPECTED LIFE		S	S		+	-	S	-	-	-	-	-
COST TO MFG		-	S		-	+	+	-	-	S	-	-
EASY to INSTALL		S	S		S	S	+	-	S	-	-	-
SHELF LIFE		S	S		S	S	-	-	S	S	S	S
Total +		0	2		8	3	5	3	0	2	2	2
Total −		6	9		1	9	7	12	11	8	13	13

Adapted from S. Pugh "Concept Selection - A Method That Works" International Conference on Engineering Design. 1981

Figure 7-3

The alternatives are presented in sketches that serve as a quick reminder of the concepts. Each picture contains the same level of detail so that decisions are not influenced by technological sophistication.

In the next step, the team compares the alternatives using performance measures and target values (the same analysis presented in Chapter 5, the Demanded Quality vs. Performance Measure Matrix).

Analysis of Design Options

Figure 7-3 shows one way of comparing different designs for the car horn. The consumer's demanded quality could be Horn Can be Heard. The related performance measures are Sound Power (with a target of 106-125 dB) and Frequency Range (with a target of 2000-3000 Hz). Only the first two rows are presented as performance measures with targets.

The rows for the comparisons among design options must include all the perspectives that are relevant to the design. This is the only stage in QFD that incorporates everything into a single analysis. Accordingly, Figure 7-3 includes failure modes, design specifications, performance measures, safety and reliability issues. As the fourteenth row indicates, this is an analysis where cost becomes relevant. In addition, manufacturability and ease of disposal are also important.

When teams make comparisons among designs, they use the existing design as the reference. The reference is considered world class or the best that is available.

Each alternative is compared to this reference. Teams utilize a three-point scale by asking the question, "Is the alternative's performance superior, the same or poorer than the reference's performance?" Evaluations of +, S and - are used to represent superior, the same or inferior performance, respectively. AHP can be used for these comparisons.

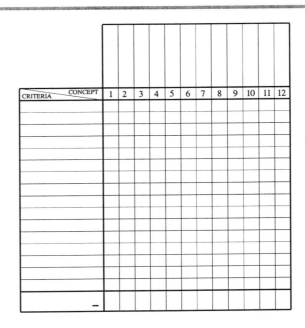

Frequently new ideas are created during discussions. Add them to the list.

The number of pluses and minuses is found for each alternative. In our example, the fifth alternative seems to be the best because there are seven more pluses than minuses; it has the highest "score" of all of the considered design options.

Concept 4 did not work; therefore, its entries were covered in column four. The team must create additional documentation explaining the technical aspects of each alternative in greater detail.

At this point you may have learned enough about design selection and choose to end your use of QFD. One organization made this choice after three trials. The team members were investigating seven different bonding formulations. After they evaluated three of the seven alternatives, they structured their data in the format shown in Figure 7-3. The newly created information showed that one of the three formulations was already better than necessary. This discovery eliminated six months of testing the remaining four formulations.

Workshop 7-2

Use the Pugh worksheet in the back of this workbook.

Enter only eight criteria in the rows. These could include performance measures, failure modes, design features, reliability issues and cost.

Enter only five concepts.

Complete the first phase of Pugh analysis.

Which alternative appears to be the best?

Figure 7-4

Concepts

Performance	Importance	The POST-IT™ Best Industry	Concept 1	Concept 2
Write on with 12 pens	32		—	—
Sheet removal	12		—	S
Shear various surfaces	26		S	S
Peel energy	26		+	—
Backboard stability	2			
Packaging test	2			
Number of Plus			1	0
Number of Minus			2	2
Weighted Plus			26	0
Weighted Minus			-44	-58

Figure 7-5

It is time to and review your current understanding.

At this point in Pugh's Selection, teams can evaluate their concept in two different ways. First, they can assign equal importance to all of the criteria and look for the best overall alternative, as shown on the previous page. In the second method, team members can ask, "What if some of the criteria are more important than others?" This important question can be answered systematically by finding a weighted total of pluses and minuses before finding a difference, as shown in Figure 7-5. The importance figures for the rows in Figure 7-5 come from Figure 5-25 in Chapter 5.

Returning to the easel pad example, if only the performance measures found in Chapter 5 are represented in the criteria set, then a weighted count could be found by adding up the weights for the pluses and adding up the weights for the minuses. Figure 7-5 has the four most important performance measures with the decimal multiplied by 100. If the importance of Peel Energy were several orders of magnitude larger than the other performance measures, the weighted approach would indicate that Concept 1 is best.

ELEV. 3000 M

Two performance measures were crossed out because they had nothing to do with the evaluation. The four most important performance measures pertain to the sheets for the easel pad. The other two performance measures relate to the easel pad's backing, which is a different design issue.

Figures 7-5 and 7-6 include criteria that involve a mixture of evaluation dimensions. AHP can be used to rank the criteria so that the design team can focus on the most important ones. First, the major groups of performance measures, cost, reliability and safety are ranked. Using the process shown in Chapters 3 and 4 (AHP and affinity trees), you can calculate an overall importance for the elements of the groups.

Creating New Concepts

Before continuing, it is necessary to repeat the analysis with concept 5 as the new reference (Figure 7-6). Column four has been eliminated. It is important to capture any small gems contained in the other designs. For example, for the criterion Temperature Resistance, Concepts 2, 5, 9 and 10 were all the same when compared to the original reference in Figure 7-3. But when Concepts 2, 9 and 10 are compared with Reference 5, Concept 2 is considered better than 5. Is it worthwhile to study Concept 2 and generate several variations on the basic theme of Concept 5?

CRITERIA \ CONCEPT	1	2	3	5 (Reference)	6	7	8	9	10	11	12	New Concept
106-125 dB	-	-	-		-	+	S	-	-	-	-	
2000 - 3000 Hz	-	-	-		-	-	S	-	-	-	-	
NO CORROSION	S	-	-		-	-	+	-	+	-	-	
SHOCK RESIST	S	-	-		-	+	-	-	-	-	-	
TEMP RESIST	S	+	-		-	-	-	S	S	-	-	
RESPONSE TIME	-	-	-		-	-	-	-	-	-	-	
COMPLEXITY	S	-	+		+	+	-	-	-	+	+	
POWER NEED	-	-	-		-	-	+	-	-	-	-	
EASY to MAINTAIN	-	-	+		-	-	-	-	-	-	-	
WEIGHT	-	-	-		-	-	-	S	-	-	+	
SIZE	S	-	-		-	-	-	-	-	-	-	
NUMBER of PARTS	-	-	-		-	-	-	-	-	-	-	
EXPECTED LIFE	-	-	-		-	-	-	-	-	-	-	
COST TO MFG	+	-	+		+	+	+	+	+	+	+	
EASY to INSTALL	S	+	-		+	+	-	+	+	-	-	
SHELF LIFE	S	S	S		S	-	-	-	-	+	S	
Total +	1	2	3		3	5	3	2	3	3	3	
Total −	5	13	12		12	11	11	12	12	13	12	

Adapted from S. Pugh "Concept Selection - A Method That Works" International Conference on Engineering Design, 1981

Figure 7-6

Pugh's Selection is an exciting procedure that produces creative ideas and innovative solutions. In 1985, a manufacturer of air bag covering for cars used QFD. The team initially generated three ideas. After the first cycle, four new concepts were generated. Two of them were sent to field testing and the manufacturer issued seven patent applications. During the field testing, employees were curious about the process the design team used to create their unusual designs. This organization now has 70% of the market.

To generate additional ideas, begin a second cycle using best concept from cycle one. Figure 7-6 has 5 as the new reference for the next series of comparisons. This process would continue until there was no change in the reference. The final reference is used as a base for new concepts.

Adapting an idea from one of the other designs to modify the reference stimulates new concept development. It is common for a group to generate several fresh, unrelated ideas during discussion. These should be added to the list. The energy of the team is frequently very high as team members escape from the confines of their traditional thinking patterns.

Do any of the columns that have + or S suggest alternative designs for the current best? If there are several current best, this procedure can be used for all of them.

Every alternative design except 2 offers an improvement on the cost to manufacture. This should create several new concepts. Another cycle of analysis is conducted with the additional concepts. The process stops when there is no further improvement or new ideas.

This approach for evaluation and generation of new ideas can be used at either the system or component levels. Pugh's Selection is often used as a "stand-alone" tool in the design process. For such an application, the team would list either the demanded qualities or the performance measures for criteria and use these in the diagram's rows.

Workshop 7-3

If the original reference was the best after the analysis, skip the next line and integrate the good ideas from the reference into your design concept.

Select the best concept from Workshop 7-2 and compare all the concepts with the new concept.

If a new concept has replaced the original reference as the current best, generate at least one new idea and document your decision.

Figure 7-7

Once the team decides on the system design, development effort concentrates on the important performance measures with the greatest gap between the target and the current design. In some instances, a team will use a classically designed experiment or Taguchi's Robust Design. This testing often yields valuable information about how certain manufacturing conditions will affect the design. Eventually, output of the design process is the specification for manufacturing to satisfy the customer. Sometimes the output may contain a blueprint for a new manufacturing process.

You may find this a good place to stop and return to your traditional ways.

It is time to

and review your current understanding.

Chapter 8 offers several exciting ways to innovate systematically. The traditional brainstorming can be supplemented by using the Theory of Inventive Problem Solving (TRIZ).

Moving on to a new way of thinking!

Chapter 8

Theory of Inventive Problem Solving

TRIZ

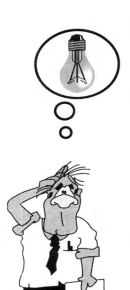

Upon completing this chapter, you will be able to:

♦ Resolve technical contradictions.

♦ Resolve physical contradictions.

♦ Utilize a standard line of product evolution.

♦ Understand Ideality.

LEVEL 1: Apparent (no invention).   Established solutions.   Well-known and readily accessible.

LEVEL 2: Improvement.   Small improvement of an existing system, usually with some compromise.

LEVEL 3: Invention inside paradigm.   Essential improvement of an existing system.

LEVEL 4: Invention outside paradigm.   A concept for a new generation of an existing system based on changing the principle of performing the primary function.

LEVEL 5: Discovery.   Pioneer invention of an essentially new system.

Figure 8-1

The History of TRIZ

TRIZ is the Russian acronym for

Теория Решения Изобретательских Задач.

The English translation, Theory of Inventive Problem Solving, produces the acronym TIPS. Operationally, some organizations are referring to TRIZ as Systematic Innovation. Any searches on the Word Wide Web should use all three names. TRIZ increases innovation and eliminates conflicts between the performance measures identified in the House of Quality in Chapter 5. TRIZ is also a powerful way to generate alternative concepts for the columns of the Pugh matrix in Chapter 7. TRIZ can be used to improve the process designs required in Chapter 9.

The creator of TRIZ, Genrich Altshuller, was a patent investigator for the Russian Navy in 1946. He saw his job as an opportunity to help inventors find creative solutions to technical problems. First he turned to the insights of psychologists to find ways to unlock inventors' creativity. Ultimately, these investigations into the behavioral sciences were not as productive as the insights he gleaned from reviewing thousands of patents. Through his work as a patent investigator, Altshuller identified patterns frequently used in innovative patents. These patents provided solutions to contradictions, and the solutions often represented one point along repeatable lines of evolution.   Altshuller's discovery replaced the sudden eureka of the stereotypic scientist with (1) regularities in design evolution, (2) the concept of Ideality and (3) the 40 principles used in most innovative patents. These three elements of TRIZ represent a fruitful mindset for innovation. There are several additional methods that fall under the umbrella of TRIZ, which are explained in my TRIZ book.

Altshuller's results attracted professionals from many disciplines who adopted and expanded his methodology. TRIZ practitioners now estimate that 1.5 million patents worldwide have been reviewed to identify patterns and regularities that have contributed to further refinement of TRIZ. Hundreds of technical papers and many books on the subject have been published. Altshuller authored 14 books alone and several with Boris Zlotin and Alla Zusman. Only two of Altshuller's books have been translated into English. This chapter of *Step-by-Step QFD* is the result of many hours with Zlotin and Zusman, chief TRIZ engineers for Ideation International, Inc.

Degree of Inventiveness

Altshuller defined an inventive problem as a problem that contains at least one contradiction. He defined a contradiction as a situation where an attempt to improve one feature of the system detracts from another feature. After reviewing 200,000 patent abstracts, Altshuller selected only 40,000 as representative of inventive solutions while the remainder involved direct improvements. Altshuller refined the different degrees of inventiveness present in the technical solutions by segmenting them into five levels of solution inventiveness (Figure 8-1).

Level 1 represents solutions of routine design problems obtained using commonplace methods within the particular specialty. The existing design system is not changed, although particular features may be enhanced or strengthened. An example of such a solution is an increase in the thickness of walls to allow for greater insulation in homes.

Level 2 leaves the existing system unchanged with or without the addition of new materials. This level always includes the introduction of new features that lead to definite improvements. One example is the inclusion of a mirror in a welder's mask to focus the light of the arc on areas needing greater visibility.

Level 3 constitutes a significant improvement of an existing system. Replacing the standard transmission of a car with an automatic transmission is one example. These inventions usually involve technology integral to other industries but not widely known within the industry in which the inventive problem arose. The resulting solution causes a paradigm shift within the industry. A Level 3 innovation lies outside the range of accepted ideas and principles of that industry.

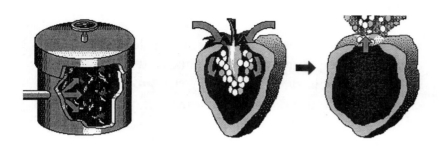

Adapted from Ideation International Software 1996

Figure 8-2

Level 4 solutions are found "not in technology but in science" that is, through the utilization of previously little-known physical effects and phenomena. Use of materials with thermal memory (shape memory metals) is one example.

Level 5 solutions are beyond the limits of contemporary scientific knowledge. This type of solution occurs when a new phenomenon is discovered and applied to the inventive problem. Level 5 solutions create new systems and industries such as lasers and transistors.

In *And Suddenly the Inventor Appeared* (pg. 87), Altshuller writes: "If one chooses to develop a completely new technical system when the old one is not exhausted in its development, the road to success and acceptance by society is very harsh and long. A task that is far ahead of its time is not easy to solve. And the most difficult task is to prove that a new system is possible and necessary." The inventor must be cautious because designs that are too advanced may not be accepted by the public. Introducing several incremental improvements is a better strategy. Radar was a new device during World War II. This invention radically increased a submarine crew's awareness of approaching aircraft. Captains of one country refused to use the radar after installation. They argued that because they became aware of twice as many planes, the radar must be attracting the planes. Today their reaction may seem strange, but it exemplifies people's initial resistance to technological innovation. For example, computer programmers initially saw no value to integrating a CRT screen with the computer.

Inventions involving levels 1, 2 and 3 are usually transferable from one discipline to another. This means that 95% of the inventive problems in a particular field have already been solved in another area.

Patterns of Inventions
While searching the patent fund, Altshuller recognized that the same fundamental problem (contradiction) had been addressed by a number of inventions in different

areas of technology. He also observed that the same fundamental solutions were used over and over again, and their implementations were often separated by many years. Let us explore several inventions.

An organization producing artificial diamonds splits the crystals at the fracture to have usable diamonds. Unfortunately, this process often results in new fractures. The process improvement team composed of engineers would not be inclined to look at agricultural patents for possible ideas. If they did, they would find:

Invention 1. Sweet Pepper Canning Method

Before canning sweet peppers, the stalk and seeds must be separated from the pod. This was done manually. Automation was difficult because the pods were not uniform in shape or size.

The following innovation created a new method for canning sweet peppers. The pods are placed in an airtight container in which pressure is gradually increased to 8 atmospheres. The pods shrink, which results in fracturing at the weakest point, where the pod bottom joins the stalk. Compressed air penetrates the pepper at the fractures, and the pressure inside and outside the pepper equalizes. The pressure in the container is then quickly reduced. The pod bursts at its weakest point (which has been further weakened by the fractures), and the pod bottom is ejected, taking the seeds with it (Figure 8-2).

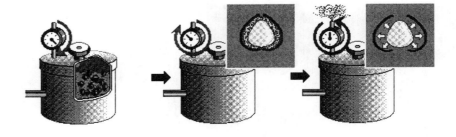

Adapted from Ideation International Software 1996

Figure 8-3

Invention 2. Shelling Cedar Nuts

The process of shelling cedar nuts is similar. They are placed underwater in a pressure cooker. Heat is applied until the pressure reaches several atmospheres. The pressure is then quickly dropped to one atmosphere. When the over-heated water penetrates the nuts, the resulting strain breaks and casts off the shells (Figure 8-3).

A similar procedure is used for shelling krill — a small crustacean that inhabits the ocean.

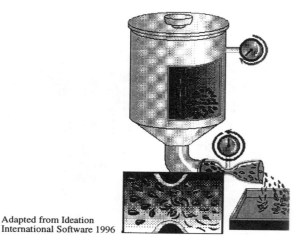

Adapted from Ideation
International Software 1996

Figure 8-4

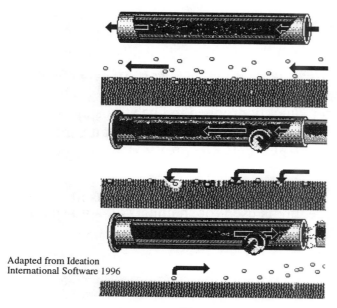

Adapted from Ideation
International Software 1996

Figure 8-5

Invention 3. Husking Sunflower Seeds

One method of husking sunflower seeds involves loading them into a sealed container, increasing the pressure inside the container, and then decreasing the pressure quickly. The air that penetrates the husks under high pressure expands as the pressure drops, thereby splitting the husks (Figure 8-4).

Invention 4. Producing Sugar Powder

A similar technique, using lower pressure, breaks sugar crystals into powder.

There are also other related solutions that could have been investigated.

Invention 5. Filter Cleaning

A filter used to treat fine-grained sand consists of a tube whose walls are coated with a porous, felt-like material. When air passes through the tube, the sand particles are trapped in the pores. Cleaning such a filter is difficult. However, the filter can be cleaned by disconnecting it from the system, sealing it, exposing it to a pressure of 5 to 10 atmospheres and then quickly reducing the pressure to normal. The sudden change in pressure forces air out of the pores, along with the sand. The sand particles are carried to the surface of the filter where they are easily removed (Figure 8-5).

These five inventions occurred in different industries at different times. If later inventors had knowledge of these earlier solutions, their tasks would have become more straightforward. Unfortunately, interdisciplinary barriers made this information unavailable. The effort to solve the diamond splitting problem could have been reduced if the researchers had read any of the above patents. Note that the principle is the same, but the design of the system and procedure differ. The diamond manufacturer eventually received a patent for the following solution.

Invention 6. Splitting Imperfect Crystals

The crystals are placed in a thick-walled, air-tight vessel. The pressure in the vessel is increased to several thousand atmospheres and then quickly returned to normal. This sudden change in pressure causes the air in the fractures to break the crystals.

Altshuller reasoned that knowledge about inventions may be extracted, compiled and generalized to enable easy access by an inventor to any area. For example, all six inventions mentioned above may be described in the following general way.

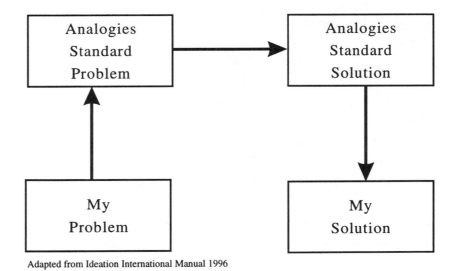

Adapted from Ideation International Manual 1996

Figure 8-6

Place a certain amount of pepper, seeds, crystals, etc. into an air-tight container, gradually apply increased pressure, then drop the pressure quickly. The sudden pressure difference creates an explosion and splits the product in the desired manner. This concept is one method for causing explosions in the TRIZ methodology.

The general knowledge generated by solutions like these can be organized and used as shown in Figure 8-6. Inventors should match their problems with similar standard problems. Then possible standard solutions associated with the standard problem are applied to the specific problem. Via this process, TRIZ accumulates innovation experience and provides access to the most effective solutions independent of industry.

Normal Problem-Solving Process

Ideation International's TRIZ training manual states: "Humans solve problems by analogic thinking. We try to relate the problem confronting us to some standard class of problems (analogs) with which we are familiar, and for which there exists a solution. If we can draw the right analogy, we can arrive at the right solution. Our knowledge of analogous problems is the result of our educational, professional, and life experiences." But what if we have never encountered a problem analogous to the one we face? The following 40 principles in Figure 8-8 answer this question by expanding our knowledge base of analogous problems (Figure 8-6).

Many problem solvers try going directly from problem to solution through trial and error. Looking at an analogous standard problem and its associated standard solution is a more efficient approach. The TRIZ methodology opens up the world patent base for identifying possible principles. This chapter offers only a few of the components of classical TRIZ. By structuring the design problem as a contradiction, the problem solver may fit a problem into the structure of the TRIZ contradiction table. The table offers several frequently used principles to solve analogous problems. The problem solver's creativity is now exercised to transfer the analogies into a useful format.

Technical Contradictions

Altshuller's definition of an inventive problem as a problem that contains at least one contradiction was based upon his observations of numerous patents. The 40,000 inventive patents he examined involved only 39 parameters. As parameter A of a technological system is improved, then parameter B deteriorates. For example, if a car accelerates faster, the fuel economy decreases. Likewise, when a product is made stronger, then it becomes heavier. The 39 parameters are listed in Figure 8-7.

Conflicting requirements are often satisfied by tradeoffs, or a compromised design.

As a result of Altshuller's work, each scientist does not have to look at all patents from all disciplines. Instead the researcher need only look at the contradictions and the associated principles to find ideas to solve a problem. These 40 principles are only a small part of classical TRIZ and an even smaller part of modern TRIZ.

Using the contradiction matrix offers several ideas for designing around a conflict. Innovations using the 40 principles eliminate the conflict produced by one or more parameters deteriorating as another is improved. These 40 principles are listed in Figure 8-8. For a detailed description of the 40 principles, see Appendix B.

Altshuller's Parameters

1.	Weight of moving object	21.	Power
2.	Weight of non-moving object	22.	Waste of energy
3.	Length of moving object	23.	Waste of substance
4.	Length of non-moving object	24.	Loss of information
5.	Area of moving object	25.	Waste of time
6.	Area of non-moving object	26.	Amount of substance
7.	Volume of moving object	27.	Reliability
8.	Volume of non-moving object	28.	Accuracy of measurement
9.	Speed	29.	Accuracy of manufacturing
10.	Force	30.	Harmful factors acting on object
11.	Tension, pressure	31.	Harmful side effects
12.	Shape	32.	Manufacturability
13.	Stability of object	33.	Convenience of use
14.	Strength	34.	Repairability
15.	Durability of moving object	35.	Adaptability
16.	Durability of non-moving object	36.	Complexity of device
17.	Temperature	37.	Complexity of control
18.	Brightness	38.	Level of automation
19.	Energy spent by moving object	39.	Productivity
20.	Energy spent by non-moving object		

Figure 8-7

40 Inventive Principles

1. Segmentation
2. Extraction
3. Local quality
4. Asymmetry
5. Combining
6. Universality
7. Nesting
8. Counterweight
9. Prior counteraction
10. Prior action
11. Cushion in advance
12. Equipotentiality
13. Inversion
14. Spheroidality
15. Dynamicity
16. Partial or overdone action
17. Moving to a new dimension
18. Mechanical vibration
19. Periodic action
20. Continuity of useful action
21. Rushing through

22. Convert harm into benefit
23. Feedback
24. Mediator
25. Self-service
26. Copying
27. An inexpensive short-lived object instead of an expensive durable one
28. Replacement of a mechanical system
29. Use a pneumatic or hydraulic construction
30. Flexible film or thin membranes
31. Use of porous material
32. Changing the color
33. Homogeneity
34. Rejecting and regenerating parts
35. Transformation of physical and chemical states of an object
36. Phase transition
37. Thermal expansion
38. Use strong oxidizers
39. Inert environment
40. Composite materials

Figure 8-8

It is not unusual for several of the principles to offer ideas. The first principle, Segmentation, underlies the design of a device for hanging sheets of paper on a wall. The device has rollers in a channel. Having several independent rotating rolls allows sheets of paper to be added or removed along the holder. A sheet of paper is positioned between the back of the device and the internal roller. Gravity provides the force that holds the paper in position; the heavier the paper, the higher the clamping force. This is a self-controlling system (Figure 8-9). Try this process in Workshop 8-1 in Figure 8-10.

The range of application of the principles is amazing. These devices even work in business environments. The 40 principles are so powerful that just looking at the list often stimulates several new ideas.

Contradiction Table

From the 40,000 patents, Altshuller selected several principles for each combination of engineering parameters. Not all of the combinations have recommended principles for conflict resolution because there was no patent to solve that particular contradiction. The parameter needing improvement is selected in the table rows of Appendix B. The recommended principles are found at the row's intersection with the column representing the degraded parameter. Typically, several technical contradictions exist and all should be investigated.

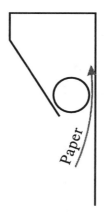

Figure 8-9

The formulation of the problem and the direction in which to focus development are the most significant processes in innovation. The creativity of the team/individual is still needed to design the appropriate system to utilize the identified principles.

Workshop 8-1

Pick a random number between 1 and 30. Find this number in Figure 8-8. Consider this principle and the nine principles which follow it. Select a product from a magazine or from your organization. Which one of these principles has been or could be used to improve the design?

Figure 8-10

Contradiction Table

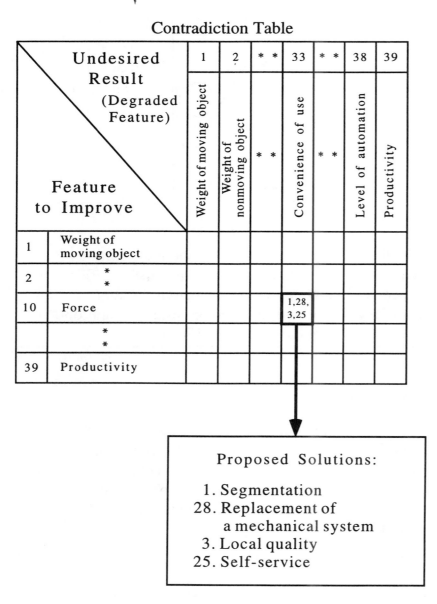

Undesired Result (Degraded Feature) / Feature to Improve	1 Weight of moving object	2 Weight of nonmoving object	* *	33 Convenience of use	* *	38 Level of automation	39 Productivity
1 Weight of moving object							
2 * *							
10 Force				1,28, 3,25			
* *							
39 Productivity							

Proposed Solutions:

1. Segmentation
28. Replacement of a mechanical system
3. Local quality
25. Self-service

Figure 8-11

Consider a spring-loaded clip as wide as an easel pad. This design exemplifies the importance of the individual's/team's decision making. If the force holding the pad in position is increased, it becomes inconvenient to use. The intersection of these parameters, 10 in the row and 33 in the column (Figure 8-11), recommends that Inventive Principles 1, 28, 3 and 25 be considered. The complete matrix is in Appendix B. The principles are ordered according to frequency of occurrence. A complete list of the order by decreasing frequency and explanations of the principles are also in Appendix B.

Each of the principles is considered for a more comprehensive list of alternatives.

28. Replacement of a mechanical system
   a. Replace a mechanical system with an optical, acoustical or odor system.
   b. Use an electrical, magnetic or electromagnetic field or interaction with the object.
   c. Replace fields
      i. Stationary fields to moving fields
      ii. Fixed fields to those changing in time
      iii. Random fields to structured ones. Use a field in conjunction with ferromagnetic particles.

Example:
1. To increase a bond of metal coating to a thermoplastic material, the process is carried out inside an electromagnetic field to apply force to the metal.

The use of a chemical field is one extension of principle 28. Continuing with this thought becomes "using the same adhesive on the sheets for the backing." The clip is eliminated. Principle 25 (Self Service) could also suggest the same idea.

A gravitational system could replace the mechanical system. This suggests using a bigger version of the gravity-operated clip shown in Figure 8-9.

If none of the principles suggests a new design, other alternative contradictions are formulated.

The contradiction table and other tools presented in this chapter can play a significant role in supporting systematic innovation. Try using the contradiction table in Workshop 8-2 in Figure 8-12 to generate new ideas for your project. Not every problem can be made to fit the contradiction matrix. To overcome this drawback, Boris Zlotin and Alla Zusman of Ideation International have extended Altshuller's work by integrating the various tools into a coherent system called Ideation Methodology. They integrated the different approaches developed by Altshuller into the TRIZ/Ideation Methodology. This integrated and comprehensive approach is supported by a software package called the Innovation Work Bench^{TM}. For more information call Ideation International at 313-353-1313 or Responsible Management at 603- 659-5186

Workshop 8-2

Formulate your problem as a contradiction between 2 of the 39 parameters. Develop a design alternative using the contradiction table in Appendix B.

Figure 8-12

Physical Contradictions

Physical contradictions are the simultaneous occurrence of one state of existence along with an opposing state of existence. Something must be both hot and cold. These contradictions are resolved by separating the phenomena. Only four separation principles are presented in this chapter. They will be used to develop alternative designs for eyeglasses to see near and far. Each separation principle should be investigated because you do not know which one will lead to the most significant breakthrough.

Separation in space: Two different lenses or bifocals.

Separation in time: Two pairs of glasses, changing back and forth as the need arises.

Separation between parts and the whole: A flat sheet of plastic with rings cut for magnification. This design can be purchased in book stores for reading small print.

Separation upon condition: The lenses are replaced with a self-focusing camera lens.

Workshop 8-3

Identify the relevant physical contradiction for your project.

What design opportunities do the separation principles provide for your project?

Figure 8-13

Patterns of Product Evolution

Altshuller noticed consistent patterns of product design evolution. Subsequent research has resulted in 250 patterns and subpatterns of evolution. Nine lines of evaluation are presented and two are discussed in detail.

1. Stages of Evolution: A technological system evolves through periods of childhood, growth, maturity and decline. Graphically this is the traditional S-curve common to many processes.

2. Evolution Toward Increased Ideality: Technological systems evolve toward more efficient designs with more functions and fewer undesired effects.

3. Nonuniform Development of System Elements: Subsystems of technological systems evolve nonuniformly. Design teams commonly do not work on the performance bottleneck. All the efforts to improve the speed of the airplane after the Wright brothers' flight were concentrated on the engine. A peak velocity of 140 mph was not surpassed until the wood biplane became a metal mono-wing.

4. Evolution Toward Increased Dynamism and Controllability: Technological systems evolve toward increased dynamism and controllability. Today's camera focusing system has evolved from a fixed focus to a manual focus to an automatic focus.

5. Increased Complexity then Simplification (Reduction): Technological systems evolve first toward complexity, then toward simplicity. The initial designs of many farm implements produced mechanical motion that simulated human action. The handheld bull rake became the horse-pulled dump rake with the farmer releasing the piles. Then the dump rake was replaced with the wheel-driven side delivery rake which makes a continuous windrow. Finally, the simple pinwheel rake has self-driven fingers propelled by the forward motion of the rake.

6. Evolution Toward Micro Level and Increased Use of Energy Fields: The energy fields can be mechanical, chemical, electrical, magnetic, acoustic, thermal, etc. If the project is to design a system for making little rocks out of large rocks, the designs will be very different for each field.

7. Evolution Toward Decreased Human Involvement: Systems develop to perform tedious functions that free people to do intellectual work or to have time for recreation.

Mono- to Bi- or Poly-systems

The transition from a mono-system to a bi- or poly-system is one of the sublines of evolution in pattern 5. There are many products which initially offered a new feature and then offered several variations of the new feature. The first ball-point pens had one blue ink cartridge, but subsequent designs offered writing with three different colors. Our easel pad example is used to trace three possible developments from mono- to poly-systems. Erasing ink marks on easel pads is not possible. There are at least two ways around this shortcoming: use an erasable marking medium or cover the marks. Masking tape is the common method of covering unwanted marks. Another option is to use paper to cover the marks. Either gluing or taping the sheet holds it in position. Whiteout tape from a dispenser supplies variable length strips with adhesive to cover marks on standard writing paper. The dispenser can be purchased in any office supply store.

Making a bigger dispenser for covering large writing could be considered an extension of this latter solution. Several rolls with different widths of white tape on a single dispenser would be convenient. This is evolution from a mono- to poly-homogeneous function is

represented by the first column in Figure 8-14. A dispenser with different colors, textures and materials offers the same function but in different formats. This is a single-function, homogeneous, shifted application of the poly-homogeneous function represented by the second column. A dispenser which includes a stapler and knife includes different functions within the same system. This design is a multi-function, heterogeneous, direct-function system represented by the third column.

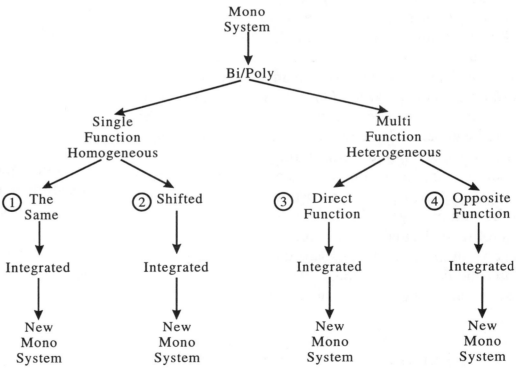

Figure 8-14

If the functions and options can be integrated, then a new mono-system has been created. The first duplication machines only duplicated. Current machines print on both sides and staple or bind the document. These machines are new mono-systems. This evolution process can be used to design tomorrow's product today. If you have a mono-system today, what are the likely bi/poly-systems of tomorrow?

Ideality

The Ideal system performs a required function without actually existing. The function is often performed using existing resources. A standard lightweight backpacking stove fueled by white gasoline is one example. The stove works when the gasoline is in a gaseous state. The liquid is first pressurized. Once the liquid is preheated to a gas, the process is self-sustaining by the heat transfer from the cooking flame down the brass construction. The pressurized liquid passes through the hot brass structure. Another example preheats air for combustion by means of the heat of combustion. Ideality is defined as the sum of a system's useful functions divided by the sum of its undesired effects. Undesired effects are also the cost of the system, the space it occupies, the noise it emits and the energy it consumes. This qualitative assessment in Figure 8-15 can be improved by any

combination of increasing the numerator and decreasing the denominator. It can also be improved by increasing the numerator at a faster rate than the denominator.

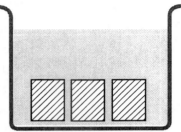

The best example involves comparing the resistance of different alloys to an acid. Several alloy specimens are placed in a closed acid-filled container. At a predetermined time, the container is opened. The effect of the acid on the specimens is measured. Unfortunately, the acid damages the container walls. The walls can be coated with glass or some other acid-resistant material, but this solution is costly. The Ideal design has a specimen exposed to the acid without requiring the use of a container. The problem is transformed to finding a way to keep the acid in contact with the specimen without a container.

Some of the resources available are the specimen, air, gravity, adhesion, etc. The solution is now obvious: make the container out of the specimen.

$$\text{Ideality} = \frac{\text{All Useful Effects}}{\text{All Harmful Effects}}$$

Figure 8-15

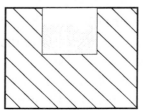

TRIZ provides two general approaches for achieving close to Ideal solutions (which increase the ratio of useful function divided by harmful function):

1. Use of resources:

A resource is any substance (including waste) available in the system or its environment which has the functional or technological ability to jointly perform additional functions. Some examples of resources are energy reserves, free time, unoccupied space, information, etc.

2. Use of physical, chemical, geometrical and other effects:

Often a complex system can be replaced with a simple one if a physical, chemical or geometrical effect is used. For example, reinforcing rods are stretched before pouring concrete during the manufacture of prestressed concrete slabs. Instead of a hydraulic system, the rods are heated and then expand by themselves by taking advantage of the coefficient of thermal expansion. The rods are then clamped into position and allowed to cool. Natural phenomena are free resources.

During colonial days in North America, granite was split in the winter by placing water in drilled holes. The increase in volume when water changes phase from liquid to solid provides the force necessary to split rocks.

There are over 250 physical effects, such as using thermal expansion, for precision adjustment. There are over 120 chemical effects, such a etching, to remove material. There are over 50 geometrical effects, such as using a mobius strip, for sanding.

Returning to our easel pad example, we see that the Ideal clip does not exist but the pad hangs. The answer is the pad itself. Each sheet is repositionable. The sheet adhesive can be used on the back of the pad. The pad can now be placed on an easel or a wall.

The insights of Ideality can be experienced in Workshop 8-4 in Figure 8-16.

Workshop 8-4

Use the concept of Ideality, reducing undesired effects and utilizing resources, to improve your product.

Figure 8-16

The TRIZ Toolbox

The complete TRIZ toolbox includes the following:

Problem Formulation

ARIZ: an algorithm for inventive problem solving
includes:
Technical Contradictions
"Ideal" final result
Physical Contradictions
"Smart Little People" model
Substance-Field Analysis

Patterns/Lines of Technological Evolution

40 Inventive Principles

Separation Principles

76 Standard Solutions to Substance Fields Models

Use of Resources

Use of Physical, Chemical and Geometrical Effects

Selected Innovation Examples

Directed Product Evolution

Anticipatory Failure Determination

The content of this chapter is an adaptation of Ideation International training material and other resources.

This material brings the user another step closer to increasing customer satisfaction.

It is time to  and review your understanding.

For those choosing to reach the final peak of our QFD journey, Chapter 9 takes a close look at ways to use QFD to improve your manufacturing process.

The journey is nearly over!

Chapter 9

Voice of Customer to Manufacturing

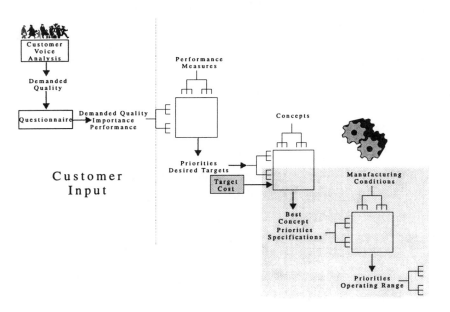

Step-by-Step QFD
for
Product Design

Customer
Input

Upon completing this chapter, you will be able to:

♦ Link the customer to manufacturing.

♦ Identify the important manufacturing conditions.

♦ Develop a manufacturing database.

EASEL PAD
TOTAL PROCESS FLOW

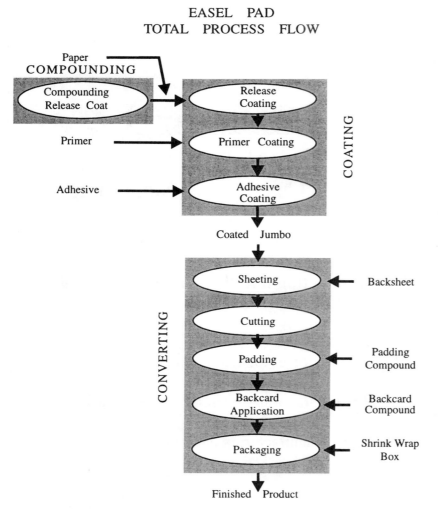

Figure 9-1

Linking the Customer to Manufacturing

The main purpose of this chapter is to show the link from the customer to manufacturing. Manufacturing deployment in QFD creates an audit trail that manufacturing process engineers use to trace their decisions back to the customer's verbatim response. The material that follows deploys the voice of the customer into a continuous process.

Easel sheet manufacturing follows the same processes as paper manufacturing, with the addition of an adhesive coating. The traditional manufacturing flow chart begins with compounds and ends with easel pads. The manufacturing flow chart moves in this direction because manufacturing staff think in terms of putting materials together and generating a finished product.

Manufacturing Process Flow

The shaded areas identify three major activities: compounding, coating and converting. Two aspects of the manufacturing flow chart can be seen. One is the flow of materials into the process and the other is the process operating conditions. Thus there is a need for two different analyses, one for the input and the other for the processing conditions. Focusing on both analyses provides teams with the opportunity to understand and adjust all the factors that they can control.

The analysis presented will first look at the characteristics associated with the materials entering the process.

From the manufacturing organization's perspective, the process starts with the raw material for compounding (shaded upper left corner of Figure 9-1).

The output of compounding is applied to the paper. This plus the primer and the adhesive are the inputs to the general process of coating. The output of this process is a very large roll of paper with a coating. In the industry, this is called a coated jumbo.

The coated jumbo plus the backcard, padding compound and shrink wrap box are the inputs into the converting process. The finished product is the output. Manufacturing acceptance tests are applied at this time.

Layout the process flow chart for your project.

Workshop 9-1

Develop a manufacturing / assembly process flow chart similar to the example.

Figure 9-2

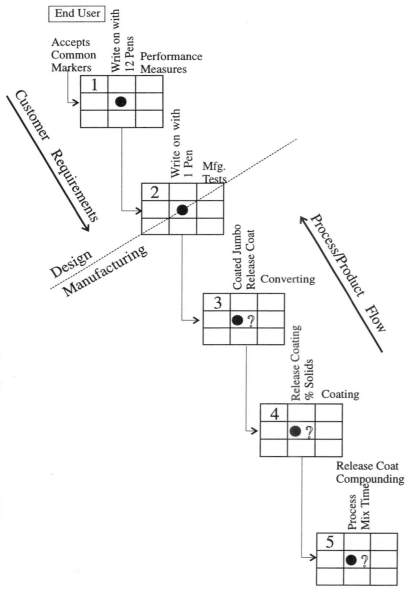

Figure 9-3

QFD vs. Manufacturing Perspective

In contrast, the QFD practitioner wants the design and manufacturing processes to be driven by the customer. Accordingly, manufacturing deployment in QFD starts with the customer and ends with the raw materials or compounds, as depicted by the arrow pointing downward from left to right in Figure 9-3. These sequence differences interfere with productive communications between design staff and manufacturing staff. Cross-functional QFD teams composed of both these staffs mimic the normal communications difficulties.

In one project, the communication problems were severe. To break the impasse, the engineers went into one room and the manufacturing and support staff went into another. The groups were to identify their needs and predict the needs of the other group. The discussion was long, heated, exciting and very productive. As you may have guessed, neither team predicted the needs of the other, but the sharing of perceptions was a start to forming effective communication patterns.

The performance measures and specifications from the earlier analysis (Chapters 4 and 5) are the input into manufacturing analysis. This is symbolized by matrix 1 in the upper left-hand corner of Figure 9-3. Pugh's Selection (Chapter 7) defined the systems design that will be manufactured. This earlier analysis clarified the priorities of these specifications and maintained the link back to the customer's verbatim response. Most manufacturing organizations have neither the time nor the equipment to evaluate all the performance measures. A subset of all performance measures becomes the manufacturing criteria for accepting a production lot. This is symbolized by matrix 2. As seen in Figure 9-3, Write on with 12 Pens is changed to Write on with 1 Pen. The second matrix in Figure 9-3 shows this transformation.

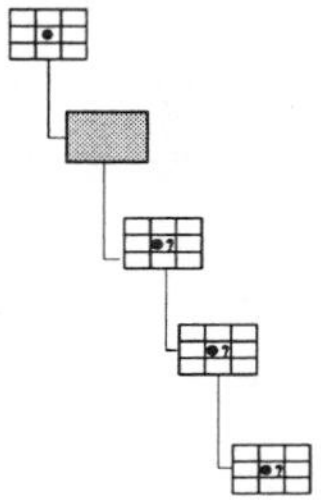

These acceptance tests must be good predictors for the original performance measures. Substitute quality characteristics are used when a direct measurement is not possible because of limited resources or physics. As mentioned earlier, placing a product in an oven with certain environmental conditions is a common accelerated age test. In some cases, the linkage between these simulations and real-world conditions is dubious. Many large corporations continue to require these tests.

Workshop 9-2

Develop the macro flow for the manufacturing process from the perspective of manufacturing, or assembly if appropriate.

Show the QFD flow through this macro flow.

Figure 9-4

Acceptance Matrix

Relationship
● Strong 9
◔ Medium 3
○ Weak 1
None 0

		% Importance	Specification	Manufacturing Tests on Finished Product: Write on with 1 Pen	Shear to Steel Panel	Appearance	Sheet Removal	Packaging Drop Test	Stiffness
Performance Measure	Write on with 12 Pens	32	3	●					
Performance Measure	Sheet Removal	12	0				●		
Performance Measure	Shear to Various Surfaces	26	100		●				
Performance Measure	Peel Energy	26	190		○		●		
Performance Measure	Backcard Stability	2	2					●	●
Performance Measure	Packaging Test	2	P			◔		◔	
	Raw Importance			288	260	6	315	21	18
	% Importance			32	29	1	35	2	2
	Specification								

Figure 9-5

Acceptance Matrix

Once a QFD practitioner has learned the process for linking the demanded qualities to the performance measures, most of the other analyses feel very familiar.

The row inputs are the performance measures and their % Importance from the Demanded Quality vs. Performance Measure Matrix (Figure 5-25). All calculations in this chapter will use the rounded numbers from Chapter 5 for demonstration purposes. The 9, 3 and 1 are used for calculations and are represented by the filled-in circle, the 25% filled-in circle and the blank circle.

Whenever you finish any QFD analysis, ask the question:

"What is the next most important aspect of the project that needs to be understood?"

At this point, the answer is identifying a suitable set of manufacturing tests.

In a real application, the list of performance measures would be much longer. But for a first time QFD, only a subset of the more important performance measures would be used. Deciding what size subset is best involves a judgment call. Asking the following question helps teams decide what to include in the set:

"What is the purpose of doing this analysis?"

In this case, we want meaningful measurements that are compatible with manufacturing's resources. This matrix (Figure 9-5) could be called an engineer's performance measures vs. manufacturing's finished product tests matrix. The cell entries represent each test's ability to predict each performance measure. If manufacturing can perform all tests, then this transformation is not necessary. The weighted Raw Importance is calculated by multiplying the column of % Importance by the column for each test. For example, the weighted Raw Importance for Shear to Steel Panel = 26 x 9 + 26 x 1 = 260. The % Importance is calculated for each column and the appropriate target is set.

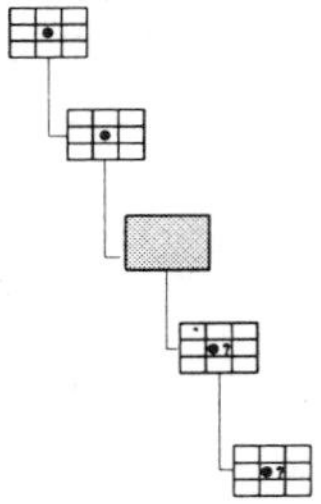

These manufacturing tests are the input into the rows of the first matrix in the macro flow (Figure 9-3). In this case it is the converting process (the third step in the QFD process and the last operation from the manufacturing perspective).

The generic structure of the matrix uses the manufacturing tests (performance measures) as the rows and the manufacturing conditions and system inputs as the columns. Normally, two different matrices are used. In this case both sets of information are placed in a single matrix to facilitate understanding of how **both** influence the substitute quality characteristics.

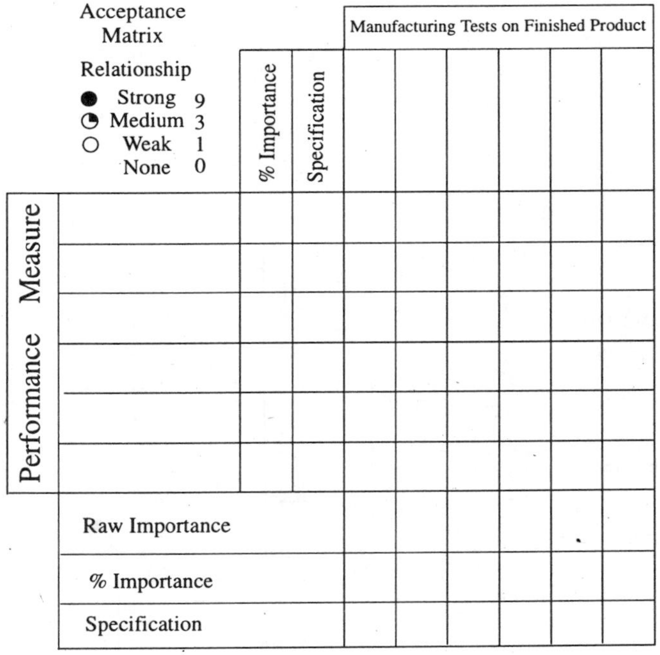

Workshop 9-3

Use the acceptance matrix worksheet from the back of the book.

Select the three most important performance measures from your previous work.

Identify the manufacturing equivalent tests for these performance measures.

Figure 9-6

Customer-Driven Manufacturing

If the customer is to drive the design process and the manufacturing process, then the deployment must start with the finished product and end with the raw material.

Accordingly, the first analysis begins with the converting process.

Notice that the converting matrix has a different structure than the previous matrices. This structure provides the answer to the question, "What is the purpose of this matrix?" The rows are the manufacturing tests from the columns in the acceptance matrix. The information in column 1 also comes from the acceptance matrix. Column 3 is an indication of the repeatability of the test used to measure the substitute quality characteristic.

Converting Process Step		% Importance of Mfg. Test 1	Specification 2	Test Method Adequacy 3	Process Capability 4	Component Characteristic Process Parameter	Process Steps Component	Input Coated Jumbo					
								Release Coat 5	Adhesion 6	Transfer 7	Stripe Width 8	Stripe Quality 9	10-20
Write on with 1 Pen	A	32	-0-	●				●					...
Shear to Steel Panel	B	29	150 gm	◖	Cpk= 1.7				◖	◖			...
Sheet Removal	C	35	40 gm	◖	Cpk= 1.2			●	●		◖	○	
	D												
Raw Importance								603	402	87	105	35	
Approved Operating Range								150 100-200	50 40-65	25 0-70	1.5 +/-.25		

Figure 9-7

Test Adequacy

● Ideal (Precision & Repeatability)

◖ Adequate (Precision & Repeatability)

○ Poor (Precision & Repeatability)

? No (Precision & Repeatability)

Column 4 contains the manufacturing capability. The Cpk index is the ratio of the number of manufacturing process standard deviations to the nearest specification divided by 3. This measures both process centeredness and variability. To further simplify the matrix, only the characteristics associated with Coated Jumbo roll of paper are shown in columns 5 through 9. The symbols are positioned for additional information.

The columns on the right-hand side under Input show three levels of detail: Input, Coated Jumbo and five specific characteristics of Coated Jumbo. This information structures the broader picture of this particular manufacturing process step. The figure below provides a close-up of the various levels of detail.

Level 1 - Input									
Level 2	Coated Jumbo					Backcard	Pad. Comp.	Back. Comp.	Shrink
Level 3	Release	Adhesion	Transfer	S. Width	S. Quality				

Column 10 contains all the columns related to all the remaining input to the manufacturing process. It represents what would be about ten additional columns in a real-world example.

The cell content now answers the question:

"If the characteristic of Coated Jumbo changed, how much is the row influenced?"

Once again, the 9, 3 and 1 are used for the calculations and are represented by the filled-in circle, the 25% filled-in circle and the blank circle. The Raw Importance for column 5 is the sum of the products of columns 1 and 5.

Workshop 9-4

Remove the worksheet in the back of the book.

Brainstorm several inputs into the process.

Select one input and identify three characteristics.

Answer the question, "If this characteristic changes, how much is the row influenced?"

Record the appropriate symbol.

Figure 9-8

This is very similar to the process used in Chapter 5 where we calculated the weighted importance for the performance measures.

The output of this analysis is the weighted importance and the operating range for the characteristics of the coated jumbo and the other input. The % Importance would normally be calculated but is not appropriate for this truncated example.

Manufacturing Database

A second set of symbols is entered in the cells to indicate the database upon which manufacturing process decisions are being made. The standard set of symbols is used with a new meaning.

Level of Knowledge

● Response Curves from Experiments

○ Harry says, "If this goes up then that goes down."

◑ History Tables

? No idea

Harry is a generic name for an operator familiar with the manufacturing system. In Figure 9-9, the empty circles to the right of cells A5, C8 and C9 indicate that only Harry understands the details of these manufacturing relationships. Having the manufacturing database in Harry's mind is very dangerous in today's organizational downsizing environment. Harry may retire early or die unexpectedly.

Converting Process Step		% Importance of Mfg. Test (1)	Specification (2)	Test Method Adequacy (3)	Process Capability (4)	Component Characteristic Process Parameter (Component Process Steps)	Input — Coated Jumbo Release Coat (5)	Adhesion (6)	Transfer (7)	Stripe Width (8)	Stripe Quality (9)	10-20
Write-On with 1 Pen	A	32	-0-	●			●○					...
Shear To Steel Panel	B	29	150 gm	◑	Cpk=1.7			◑?	◑?			...
Sheet Removal	C	35	40 gm	◑	Cpk=1.2		●◑●●			◑○○○		
	D											
Raw Importance							603	402	87	105	35	
Approved Operating Range							150 100-200	50 40-65	25 0-70	1.5 +/-.25		

Figure 9-9

Looking at the cells showing strong impact in Figure 9-9, we see that this organization has a dangerous case for row A and column 5. There is a strong relationship between the input and the manufacturing tests, but only Harry understands the relationship between Release Coat and Write on with 1 Pen.

Cell B,7 presents more of a challenge. The relationship appears to be produced by some kind of magic rather than through management's efforts.

The operating range and importance become the input into the upstream manufacturing process. Coating is the next matrix in the customer voice deployment (Figure 9-11).

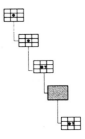

Workshop 9-5

Continue with the previous workshop by discussing the knowledge base for the relationships in your matrix.

Figure 9-10

EASEL   PAD

Coating

Process   Step

Coated Jumbo	Process Step		% Importance of Characteristic 1	Specification 2	Test Method Adequacy 3	Process Capability 4	Component Characteristic Process Parameter	Component Process Steps	Surface Paper Treatment 5	Release Coat Compounding % Solids 6	Viscosity 7	% Water 8	% Solids Adhesive 9	10
	Release Coat	A	61	150 100-200					●●◑●○	?	●○			
	Adhesion	B	41	50 40-65								●○		
	Transfer	C	9	25 0-70								○◑		
	Stripe Width	D	11	1.5 +/- .25										
	Stripe Quality	E	4											
		F												
		G												
		H												
	Raw Importance								549	549	61	183	378	
	Approved Operating Range									15%	200 cps	2%	2.5%	

Figure 9-11

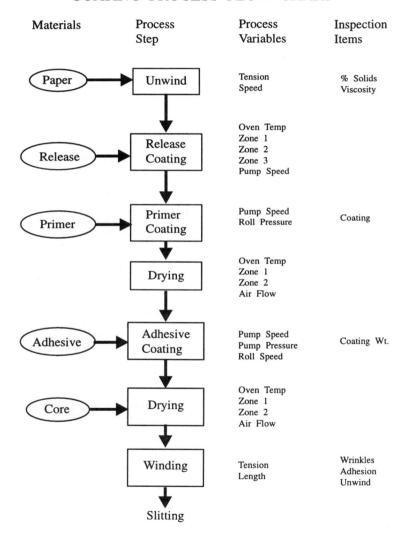

EASEL PAD
COATING PROCESS FLOW CHART

Materials	Process Step	Process Variables	Inspection Items

Paper → Unwind — Tension, Speed — % Solids, Viscosity

Release → Release Coating — Oven Temp Zone 1, Zone 2, Zone 3, Pump Speed

Primer → Primer Coating — Pump Speed, Roll Pressure — Coating

Drying — Oven Temp Zone 1, Zone 2, Air Flow

Adhesive → Adhesive Coating — Pump Speed, Pump Pressure, Roll Speed — Coating Wt.

Core → Drying — Oven Temp Zone 1, Zone 2, Air Flow

Winding — Tension, Length — Wrinkles, Adhesion, Unwind

Slitting

Figure 9-12

"What is the next most important aspect of the manufacturing process to understand?"

The coating process is. Before a coating process matrix can be added, a manufacturing process flow chart must be completed. Inspection Items are included in this flow chart (Figure 9-12). The process and analysis are the same as for the previous matrices.

A second matrix for the coating processing conditions is represented by columns 11 to 14 (Figure 9-13). Placing the input and the process matrices next to one another enables the QFD practitioner to understand how both influence the substitute quality characteristics.

Control Points and Check Points

Control points and check points are critical to QFD manufacturing deployment. Control points measure the results of a particular task or process, while check points measure the processing conditions. One check point for coating is the oven temperature in zone 1 (Figure 9-13, column 11). A control point is Coated Jumbo's Release Coat (row A). If both of these have acceptable operating values, then no actions are taken.

Once a team establishes the control points and check points for each step of manufacturing, it is easier to track the impact of processing decisions.

Tracking the Impact of a Decision

Manufacturing organizations are often encouraged to increase output as a means to improve profits. The analysis presented in this chapter offers a road map to test the consequences of any decisions to change the manufacturing process, such as increasing pump speed in column 14, in Figure 9-13.

Looking at column 14, we see a very sound database for the relationship between Pump Speed and Release Coat in row A. Only if the considered speed is outside the acceptable range is it necessary to track the consequences.

If a considered change does not force the Release Coat characteristic out of its approved operating range, there is no problem. However, if the characteristic is out of range, then QFD offers an audit trail. The release coat priority and approved operating range came from the Converting Matrix in Figure 9-9 or Figure 9-7. Since all the analyses in QFD are linked, the team does not have to dig deeply into its previous examinations to understand the consequences of a considered decision.

EASEL PAD

Coating Process Step			1 % Importance of Characteristic	2 Target Value	3 Test Method Adequacy	4 Process Capability	Component Characteristic Process Parameter (Steps Process Component)	Input						Process				
								5 Surface Paper Treatment	6 % Solids	7 Viscosity	8 % Water	9 Adhesive % Solids	10	11 Zone 1	12 Zone 2	13 Zone 3	14 Pump Speed	15 Adhesive Coat Pump Pressure
External Coated Jumbo	Release Coat	A	61	150 100-200	○			●◖	●●	○ ?		●○		●●●	○		●●	
	Adhesion	B	41	50 40-65		Cpk 1.4						●○						●○
	Transfer	C	9	25 0-70								○●						
	Stripe Width	D	11	1.5 +/-.25										●●				
	Stripe Quality	E	4															
Internal		F																
Coating	Splice Free Roll	G			●													
	Elemendorf Tear	H			●													○○
	Importance							549	549	61	183	378						
	Approved Operating Range								15%	200 cps	2%	2.5%		200			100 rpm	

Figure 9-13

The Release Coat Importance was influenced by the manufacturing acceptance tests of Write on with 1 Pen and Sheet Removal.

The Write on with 1 Pen and Sheet Removal for manufacturing came from the design team's performance measures in Figure 9-5. The Write on with 1 Pen came from Write on with 12 Pens. Sheet Removal came from Sheet Removal and Peel Energy.

These three performance measures came from the demanded qualities in the summary matrix in Figure 5-25 at the end of Chapter 5. These performance measures were the result of the consumer saying, "I want:

1. to have smear-free writing on the easel pad."

2. to be able to use the markers I bought at the local store."

3. the sheet to stay on the wall."

4. to be able to easily remove the sheet from the wall."

 Manufacturing now has a clearly defined link back to the customer.

This process becomes a powerful tool for process improvement and control. The matrices are called the Process Planning Matrices. The Importance row identifies which processing conditions should be monitored. An acceptable operating range can be determined.

Additional tools include design of experiments and Robust Designs for manufacturing. Both of these can help reduce costs, determine the best operating conditions for the desired characteristic values for the output and reduce the impact of raw material variation to create a more stable manufacturing process.

Finally, all these activities add to the manufacturing database. It is this type of information which should be behind the filled-in circle in cell A,14 of Figure 9-13.

It may be time to **STOP** and review your understanding.

After a few final comments, you will be ready to go off on your own journey and try QFD on a real project.

You may choose to take a few more steps and learn a little more about Robust Design in the next chapter.

Chapter 10

Robust Design in Manufacturing

Upon completing this chapter, you will be able to:

♦ Conduct a simple Robust Design.

Performance Measure / Demanded Quality	xxxx	Curvature	xxxx	xxxx
Stylish		◉		
Easy to Use				○
Easy to Assemble				
Carry Heavy Load				◕
Technical Benchmarking		Use different environments to find the average and σ^2		

Figure 10-1

Product Performance Measure / Operating Conditions	xxxx	Melt Temp.	xxxx	xxxx
xxxx	∘	∘	∘	∘
Curvature	∘	◉ ?	◕ ○	∘
xxxx	∘	∘	∘	∘
xxxx	∘	∘	∘	◉◉
Operating Specifications				

Figure 10-2

Robust Design

Chapter 6 presents the philosophy behind Taguchi's Robust Design and explores ways to reduce the impact of uncontrolled influences upon product performance. Since this chapter investigates a Robust Design application for manufacturing, you may want to read Chapter 6 first, if you have not already done so.

Our Robust Design application uses the easel pad example developed in the previous chapters. For this project, the easel pad team designed a long plastic clip to hold the pad and/or sheets to the easel. The required system design was the result of completing Pugh's Selection. Injection molding was the selected manufacturing process. Curvature was an important performance measure for the customer's demanded quality of Stylish Design (Figure 10-1).

Robust Design for a product or process has good quality even with uncontrolled sources of variation. Once a team selects a Robust Design, deployment into manufacturing is possible. Improving performance of a poor product design by adjusting manufacturing is not as successful as improving the design first.

QFD: Manufacturing Deployment

The first step in manufacturing deployment is making a Performance Measure vs. Operating Condition Matrix. The team identifies important control variables. Plastics manufacturers usually use melt temperature, core mold temperature, cure time, injection speed and cavity mold temperature for controlling an injection molding machine. Sources of uncontrolled variation include variation in raw materials and mold temperature drift.

The strength of the relationships between the operating conditions in the columns and the performance measures in the rows is based upon prior experience, as presented in Figure 10-2. The dot indicates that the team discussed the cell and decided that there is no relationship.

Finding the Best Operating Conditions

First a small example of baking a cake will be used to explain the procedure for a Robust Design investigation. Only the density of the cake will be considered as a measure of quality.

An ideal cake should be light. In other words, the cake should have low density. In this example, the ingredients are supplied premixed and only water is required for preparation. Only three process elements are controllable: the amount of water, the oven temperature and the mix time. The premixed material does not always have the same consistency. Two possible values for each of the processing conditions result in eight possible combinations of the controlled processing conditions (Figure 10-3). Two or three cups of water, 350 or 380 degrees for the oven temperature and five or ten minutes for mix time are the options for this example.

Control Variable / Combination	Cups of Water	Oven Temp.	Mix Time
1	2	350	5
2	2	350	10
3	2	380	5
4	2	380	10
5	3	350	5
6	3	350	10
7	3	380	5
8	3	380	10

Figure 10-3

Control Variable / Combination	Cups of Water	Oven Temp.	Mix Time
1	2	350	5
2	2	380	10
3	3	350	10
4	3	380	5

Figure 10-4

Control Variable / Combination	Cups of Water	Oven Temp.	Mix Time	Result 1	Result 2
1	2	350	5	4	3
2	2	380	10	6	8
3	3	350	10	2	2
4	3	380	5	5	6

Figure 10-5

The controlled processing conditions are commonly called control factors. Only a subset of all combinations is used for an efficient investigation. The size of the subset depends upon the number of control variables and the number of possible values for each control variable being investigated. These subset combinations are called orthogonal arrays. The array for three factors with two levels uses four combinations (Figure 10-4).

Two cakes are baked for each of the four combinations and the density is recorded (Figure 10-5). The combinations in row 1 and row 2 have two cups of water. The average of the four data for two cups of water is compared to the average of the four data for three cups of water. The two cups of water combination has an average density of (4+3+6+8)/4= 5.25 and the three cups of water combination has a density of 3.75. Three cups of water is more desirable because of the lower density. The same calculation is made for oven temperature and mix time. An average response table (Figure 10-6) summarizes the results.

Water	Avg.	Temp.	Avg.	Time	Avg.
2	5.25	350	2.75	5	4.50
3	3.75	380	6.25	10	4.50

Figure 10-6

162

The best combination is three cups of water and an oven temperature of 350 degrees. The mix time has no effect because the densities are the same for both times. However, mix time is a necessary part of the process because control variables that have no effect provide cost savings. Five minutes of mixing is less expensive than ten minutes.

A performance prediction can be made using the best combination based on the data in Figures 10-5 and 10-6. The grand average (average of the averages) of the data is (4+3+6+8+2+2+5+6)/8 = 4.5. To determine how much the average improved by selecting three cups of water, calculate the difference between the average for three cups and the grand average (3.75 - 4.5) = -0.75. The prediction equation only uses the important variables.

Predicted density = grand average
                 + improvement from water
                 + improvement from oven temp.

Predicted density = 4.5
                 + (3.75 - 4.5)
                 + (2.75 - 4.5)
                 = 2.00

Finding the confidence interval for this prediction is beyond the scope of this chapter.

Reducing Variation

The two cakes baked using each combination of variables were different. The difference in these pairs of results would increase if the first cake used batch 1 of premix and the other used batch 2 of premix. By conducting the investigation in this manner, it becomes possible to calculate which combinations of controlled variables can reduce the impact of batch-to-batch variation. This batch-to-batch variation reflects reality.

Control Variable Combination	Cups of Water	Oven Temp.	Mix Time	Batch 1	Batch 2	σ^2
1	2	350	5	4	5	0.5
2	2	380	10	6	10	16.0
3	3	350	10	2	4	0.5
4	3	380	5	5	8	4.5

Figure 10-7

The right-hand column in Figure 10-7 shows the variance for each trial. A robust process not sensitive to this uncontrolled source of variation would result in better quality with no increase in cost. Alternatively, the variation between batches can be reduced by purchasing a more expensive premix. An average variance table (Figure 10-8) facilitates selecting the best values for the control variables. The process being shown here is conceptually correct but mathematically incorrect. For the statistically minded, the log of the variance is the correct measure. The cake example is intended to promote conceptual understanding.

Water	σ^2	Temp.	σ^2	Time	σ^2
2	8.25	350	0.50	5	2.50
3	2.50	380	10.25	10	8.25

Figure 10-8

A small variance is preferred. The best value for minimum variance has 2 cups of water and 350° oven temperature, the same combination as when minimizing the density. Interestingly, the mix time influences the variation. A five minute mix time is best for reducing sensitivity to variation in batches of raw material. In this example, there is no conflict in the best combinations for minimizing both the average and the variance. A signal-to-noise ratio is normally used to incorporate the concept of the loss function in the analysis. This is also beyond the scope of this manual.

We can now apply the process used for the cake example to the easel pad clip. The 5 factors and 2 interactions used 8 different combinations instead of 128 (Figure 10-9).

Control Variable Levels Studied

Melt Temp.	500° F	535° F
Core Mold Temp.	85° F	135° F
Cure Time	150 sec	200 sec
Injection Speed	1 sec	2 sec
Cavity Mold Temp.	75° F	100° F

The Interactions

Cure Time with Cavity Mold Temp.
Cure Time with Core Mold Temp.

Eight different combinations were used to study the five variables and two interactions. The layout and results are shown in Figure 10-9. The calculation to determine which interactions are significant is beyond the scope of this material. The investigation revealed that the cure time and cavity temperature interaction was significant.

Four different combinations of what are normally uncontrolled variables were used. The melt temperature varied by ± 10 degrees during production. Two different batches of material were selected to represent possible diversity in the quality of the raw materials. The team made one clip for each combination of production conditions and uncontrolled variables (*noise*). Variation in materials and temperature did cause variation in the curve measurements.

The average responses in Figure 10-10 show that the melt temperature, core mold temperature and cure time were important for reducing the plastic clip's deviation from flatness.

	Melt Temp	Core Temp	Cure x Cavity	Cure Time	Injection	Cure x Core	Cavity Temp	1	2	1	2	Batch
								-10	+10	+10	-10	Melt Temp ▲ T
1	500	85	1	150	1	1	75	77	69	61	53	
2	500	85	1	200	2	2	100	32	39	43	38	
3	500	135	2	150	1	2	100	81	74	81	123	
4	500	135	2	200	2	1	75	83	92	66	8	
5	535	85	2	150	2	1	100	96	92	91	111	
6	535	85	2	200	1	2	75	79	74	74	100	
7	535	135	1	150	2	2	75	80	94	83	127	
8	535	135	1	200	1	1	100	75	63	79	58	

Figure 10-9

Melt	Avg.	Var.	Core	Avg.	Var.	Cure	Avg.	Var.	Inject	Avg.	Var.	Cavity	Avg.	Var.
500°F	68.7	385.10	85°F	70.6	68.74	150 sec	87.1	217.11	1 sec	76.3	161.27	75°F	81.3	402.72
535°F	86.0	150.03	135°F	84.2	466.39	200 sec	67.9	318.02	2 sec	78.4	373.86	135°F	73.5	132.41
Difference	17.3	235.07	Difference	13.6	397.65	Difference	19.2	100.91	Difference	2.1	212.59	Difference	7.8	270.31

Figure 10-10

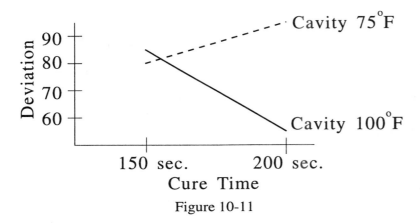

Figure 10-11

The interaction between cure time and cavity temperature was also important. The best operating conditions were 500°F for the melt temperature, 85°F for the core temperature, 200 sec. for cure time and 100°F for the cavity temperature. The cavity temperature was important because of its interaction with the cure time. Figure 10-11 shows this relationship. This study revealed that no matter what degree of curvature a team has targeted, the team must consider the relationship between cure time and cavity temperature.

Dynamic Applications

A dynamic application searches for a factor that would allow convenient adjustment of the injection machine for any mold design. The ideal function has the same percent shrinkage in all directions for all dimensions.

Summary for Robust Design

Taguchi's loss function offers an improved way to accomplish technical benchmarking in the House of Quality. Quality is measured by average performance and variance. The targets for your design must also consider the average and variance. In addition, teams should evaluate the competitor's products in a variety of environments.

A robust process gives QFD teams an edge over the competition because it reduces sensitivity to uncontrolled sources of variation. Differences in components and materials are common sources of variation. Reducing the sensitivity of the process to these uncontrolled sources of variation is both a quality improvement and a cost savings. Once a team reduces the sensitivity to uncontrolled sources of variation, installing inexpensive controls and processes becomes practical.

It is time to

and review your understanding.

A few brief comments follow before you begin your role as a QFD practitioner.

Chapter 11

Admonitions

Leadership and Project Selection

Projects with good planning and organization are more likely to succeed. The organizational climate must be supportive and have quality as the cornerstone of decision making. Top management must provide leadership in the use of quality thinking. Increased competition among organizations makes it imperative that top management does more than support project teams; it must assume the critical role of leading the QFD initiative.

Many people are not interested in trying QFD because of previous experiences when they became enthusiastic but were not supported in their efforts to try something new. It is particularly frustrating to attend a course and become excited about the possible benefits, only to have this enthusiasm shut off by hearing others say, "We have been successful in the past. Why rock the boat?"

Ensuring that an organization's first QFD project is successful is one way to avoid discouraging experiences. The best way to guarantee good results is to select a project that has broad appeal; one that is simple but not trivial. For example, choose an existing product or service that offers opportunity for improvement but is not the major corporate problem. You may also want to consider the availability of marketing information, as this may limit project definition. Listing the reasons for choosing the project clarifies expectations and provides direction. Deciding how to measure the project's

effectiveness *before* it has begun can help promote the team's efforts. Data comparing project time, labor hours, customer complaints, etc. with past projects is persuasive. This approach is more effective than a presentation recommending QFD because the team feels good about its work or thinks that the QFD tools had a particular effect.

All these measures mitigate the damaging influence of corporate politics. A manager is less likely to shelve a project that is backed by convincing data. If a team continues to encounter resistance, the team members should talk to their corporate change agent and work to promote QFD as a recommended way to support the current design process. Practitioners should try to find the language and the approach that will spark management's interest in their project. If they have presented their ideas as powerfully as possible, management will support their efforts.

Designing a manufacturable, serviceable product that the consumer wants requires the participation of several organizational functions. This guarantees that the QFD team members will have adequate tools for the design process.

Group Dynamics

The realities of group dynamics make it important to put a ceiling on the QFD team's membership. A team that is too large has trouble running effective meetings. Past experience dictates that a five- to eleven-member team works best. The number is strongly influenced by the team's experiences, the corporate culture and the motives for being on the team. If someone is chosen who does not want to be on the team, asking that person not to come can prevent potential problems caused by negative attitudes. Sometimes it is politically impossible to ask a team member to stop participating. In these situations, assigning this member the role of team critic provides a positive solution. Since an effective critic must be unbiased, it follows that this critic/team member should not attend the meetings. This way the critic can review the reports with an open mind and make suggestions. Taking these proactive measures creates a team that is enthusiastic and empowered.

Given the time constraints placed upon many QFD practitioners, poor meeting attendance can become a severe problem. Unless meetings are structured and are not dependent on perfect attendance, the project is likely to die a slow death. Plan to meet on a weekly basis at the same time on a particular day for six months. Everyone can easily record this information in their appointment books and then they will lack valid excuses for not knowing when the next meeting occurs. Avoid readjusting the meeting schedule. Making a change to suit one individual rapidly escalates into a planning nightmare.

Decision making by consensus is painful and difficult work, but the results are worth every hour of discussion. QFD operates on the assumption that many heads work better than one. Individuals become more dedicated when they feel confident that their input influences a project. Accordingly, it is important that expertise, rather than position within the organization, shapes the decision-making process. Consensus decision making loses some of its power when rank takes precedence over knowledge.

Common Pitfalls and Errors

Several common mistakes can undermine a QFD project. In the 1980s the most prevalent error was attempting to work with a very large matrix. A comprehensive QFD project using 16 different matrices for a critical survival project for your organization is not a good starting place. The case studies in Chapter 1 illustrate how small, carefully defined projects can provide the best environment for new QFD practitioners.

Mixing the consumer's needs with solutions or failure modes is another common oversight. One of the biggest challenges for our professional egos is pausing and responding to the voice of the customer instead of immediately jumping to the obvious technical solution. Rapid reactions also discourage innovation because they encourage practitioners to rely on traditional methods. Mastering QFD requires tolerance of the learning curve.

The material in this course points out the many opportunities to shift the design team's efforts in and out of the QFD strategy. This is a simulation of the different levels of maturity organizations go through during several months of experience.

QFD and Your Organization's Future

Over time, your organization should increase both the depth and the breadth of QFD projects. Expanding the breadth of applications involves increasing the number of projects that use QFD. This kind of change occurs at the political, informational and strategic levels. Greater depth would include other deployments such as function deployment and failure deployment. Reading case studies of QFD projects from different industries supports your implementation of in-depth QFD. Akao's *Quality Function Deployment*, for instance, contains several cases of comprehensive QFD applications. With its broad range of case studies, this book stretches the reader's imagination. Innovation furthers the evolution of QFD and increases an organization's market share.

You have reached the end of your initial journey with QFD and a variety of future paths lie before you. These trails provide different avenues for increasing your knowledge of QFD. For example, several countries currently hold QFD symposia. These gatherings provide excellent opportunities to meet with practitioners who have gone beyond your present level of expertise. Several symposia offer tutorials for participants. Some organizations also offer public seminars.

However, travelers considering this second path should be warned that such seminars often cover the House of Quality only. Consultants can provide in-house training and help your project team investigate a variety of deployments. Each year, the QFD Institute sponsors a Master class taught by a Japanese expert. Finally, there are growing numbers of QFD resources on the Internet and the World Wide Web. Appendix C has a list of addresses.

There are many other methods that should be in an organization's QFD toolbox. The following list includes some particularly useful tools. Since some of them lie beyond the scope of this manual, you will need to learn more about them on your own.

> The Seven Quality Tools
>
> The Seven Management and Planning Tools
>
> The Seven Innovation Tools
>
> The Seven Product Planning Tools
>
> Failure Mode and Effect Analysis (FMEA)
>
> Fault Tree Analysis (FTA)
>
> Factor Analysis
>
> Customer Interviews
>
> Focus Groups
>
> Design for Manufacturing
>
> TRIZ
>
> Taguchi's Philosophy

The end of this book

    does not mean your journey is over.

There is much to learn.

You must adapt

    what has been presented

    to your own particular application.

Pursue a more comprehensive QFD.

Happy stepping,

John

A   Ω

Appendix A

QFD Cases

Going to the Gemba and a Successful QFD Project

Following are various examples of depth in the QFD process.

The case of Rehab Concepts, located in Willington, Massachusetts, was presented at the 1990 GOAL/QPC Quality Conference. The company offers physical therapy for employees who have experienced industrial accidents. When Rehab Concepts was six months old, it started a QFD project under the direction of Jim Bruer, senior vice president and chief operating officer of the parent organization. After several hours work, the group requested help with the first matrix, which had already identified ways to measure performance. I wondered if the project had direct input from patients. On examining Rehab Concepts' list of needs, I noticed that the most important was "return to work as soon as possible." Suspecting that the patients might have different needs, I persuaded Rehab Concepts' team to talk to the most relevant people, the patients. Interviewing patients for one hour revealed that their primary need was to be able "to do everything they did before," such as skiing and playing tennis. Rehab Concepts' previous analyses were driven by its source of income, the insurance company and the employer. Regaining all previous capabilities demands a different therapy protocol than merely returning to work. During subsequent discussion that resulted from this finding, the staff realized they had been operating with three different schools of thought. One approach was selected, and those therapists who favored different methods or values chose to leave Rehab Concepts. The clients were much happier with the united team. Today, Rehab Concepts' reputation is unsurpassed in the state of Massachusetts.

Jim Bruer estimates that the organization would have taken another 9 to 12 months before understanding that it had forgotten its most important customer. QFD led the organization to this conclusion in a matter of hours. Upon reflection, Mr. Bruer believes that the organization should have implemented the QFD process before offering any services. Rehab Concepts' experience demonstrates how simply using one QFD tool generates such a wealth of information that an organization may postpone further QFD activities. However, while some organizations may find that using one matrix generates enough information for their design process, most projects require a more in-depth effort. For example, Original Equipment Manufacturer (OEM) suppliers often use a four-matrix analysis. After three years of experience, a comprehensive QFD model may become appropriate. It is important to understand that each project is unique, and the team's needs shape the application of the QFD process.

One Matrix and the *Gemba* Recapture a Lost Market

Puritan-Bennett resembles Rehab Concepts, in that it only needed one matrix of customer needs to reap impressive benefits.

In 1988, this designer and manufacturer of spirometers sold its product for $4500 US. With 15% of the market, Puritan-Bennett was the industry leader. Although its sales stayed constant, its percentage of the market dropped to 11% in 1989 and to 7% in 1990. The loss of market share occurred when a new competitor offered a product with fewer features that sold for $1995 US, less than half of Puritan-Bennett's cost. Before using QFD, Puritan-Bennett employed the traditional phase review in which the engineers used their knowledge of the market, while the marketing and sales force contributed their input. Since the organization's functions were not working as a unified team, past designs using new technologies experienced cost, time and quality delays.

In one year, Puritan-Bennett used QFD to develop a new spirometer. For Puritan-Bennett, the customer demands came from the physicians' and nurses' information, supplemented by dealer and distributor input (Figure 1, page 176). The asterisks on the left and right of the customer needs indicate which current design was better, Puritan-Bennett (PB) or the competitor (WA) (Kaelin and Klein, 1992).

During the design process, there were many arguments over which engineering solution or product feature to use. Thanks to the Demanded Qualities vs. Performance Measures Matrix, decisions always favored the customer. Using this particular QFD matrix led to improved communications among the organization's different departments.

The project's major breakthrough in terms of product improvement was a modular design, the Renaissance™, which allowed the different customer segments to create their own system. The Renaissance™ met each of the customer needs. The modular prices made each segment affordable. The users attained Good Printout Quality by using their existing printers or buying a printer from Puritan-Bennett. The small spirometer and the Puritan-Bennett pneumotachs satisfied the demanded quality Easy to Hold (Kaelin and Klein, 1992). The subsequent product had additional features. This new model was also portable and priced less than the previous model ($1590 US). With a better design and reduced selling price, Puritan-Bennett could not keep up with the rising demand for its new spirometer. The new design took away the competitor's price edge, further improved Puritan-Bennett's good performance on Easy to Hold and fulfilled a need for Good Printout Quality that neither company had satisfied.

PB	Customer Needs	WA
	Product is affordable	*
	Easy to operate	*
*	Easy to clean	
	Convenient-sized output	
*	Sanitary	
*	Quick service response	
*	Provides accurate readings	
	Eliminates technician variation	
	Good printout quality	
	Reliable	
	Diagnostic information meets needs	
*	Easy to interpret	
	Fast to use	*
*	Easy to hold	
	Easy to set up the first time	*
	Easy to calibrate	
	Availability of machine supplies	*
	Good training/education	
	Sleek appearance	*
	Good printer quality	
	Low cost of repairs/service	
	Portable	*
	Effective data storage/ retrieval	
	Environmentally safe	*

Figure 1

A QFD Application with Depth

Kimberly-Clark completed a very meaningful QFD project and shared its general learning. Diane Scheurell's paper, presented at the 1993 annual QFD symposium in Novi, Michigan, revealed that the company used 16 different matrices for a comprehensive study that brought about major breakthroughs. Kimberly-Clark successfully applied QFD in the following areas:

1. Creating a new product design.
2. Using a new manufacturing process.
3. Using a new technology.
4. Manufacturing in a new plant.
5. Manufacturing with new equipment (one of a kind and some first of a kind).
6. Working with new employees (Scheurell 1992, 1993).

The team employed a series of 16 matrices. Analysis of the first matrix revealed that three other matrices were needed: a functional matrix, an appearance matrix and one for converting and packing.

Each matrix became a steppingstone to additional matrices. In the end, Kimberly-Clark began startup one month before it had originally planned and released the product two weeks early. Startup costs were below budget because the teams eliminated unnecessary steps. They also discovered equipment inadequacies early on and used statistical process control (SPC) for processing conditions to eliminate an audit program.

Examining how Kimberly-Clark achieved this success reveals the elements of a productive implementation of QFD. During the initial cross-functional team meeting, the following goals were created and approved by management:

- Deliver a quality product on time.
- Eliminate fire fighting.
- Increase communication.
- Ensure that mill operators and engineers:

  know customer needs
  understand the product and reasons for specifications
  base decisions on customer priorities.

In these early planning sessions, Kimberly-Clark was already thinking ahead about how QFD would shape its design process. After recording the previous goals, the organization created the following list of steps needed to reach its objectives:

- Develop the Process Planning Matrices.

- Develop the framework for the Production Planning Matrices.

- Assess prior equipment purchase decisions, future equipment design and purchase decisions against the Process Planning Matrices.

- Use the matrices to identify and act on important issues.

- Develop a training program in QFD for mill personnel.

- Encourage use of QFD tools by example.

One of the project's objectives was a verbal design process chart defining what analyses were needed. This allowed the team to anticipate where QFD would support the design process. Chapter 2 explored the linkage between QFD and the design process.

Another critical element was the creation of a 60-page description of the manufacturing process flow. This document included all the variables that could be controlled at each point in the process. Many of the improvements were the result of integrating this document with QFD. Chapter 8 examined how QFD uses process flow charts to improve manufacturing.

Scheurell's paper documenting this QFD project asserts that a QFD effort can flounder if the group working on it is not an empowered team (Scheurell, 1993). This group received sufficient allocation of up-front resources and time. In addition, they maintained consistent team membership which promoted a smoother decision-making process. The team also clearly defined its purpose and received management's approval and authority to do the job. Due to the project's startup nature, the team had the additional advantage of being entirely composed of managers.

Appendix B

40 TRIZ Principles and Contradiction Table

40 Innovation Principles

1. Segmentation
a. Divide an object into independent parts.
b. Make an object sectional.
c. Increase a degree of an object's segmentation.
   Example:
   1. Design sectional furniture, modular computer components or a folding wooden ruler.
   2. Garden hoses can be joined together to form any length.

2. Extraction
a. Extract (remove or separate) a "disturbing" part or property from an object.
b. Extract the only necessary part or property.
   Example:
   1. Using a tape recorder, reproduce a sound known to excite birds to scare them away from the airport. (The sound is separated from the birds.)

3. Local quality
a. Provide transition from a homogeneous structure of an object or outside environment (outside action) to a heterogeneous structure.
b. Have different parts of the object carry out different functions.
c. Place each part of the object under conditions most favorable for its operation.
   Example:
   1. To combat dust in coal mines, a fine mist of water in a conical form is applied to working parts of the drilling and loading machinery. The smaller the droplets, the greater an effect in combating dust, but fine mist hinders the work of the drill. The solution is to develop a layer of coarse mist around the cone of fine mist.
   2. Make a pencil and an eraser in one unit.

4. Asymmetry
a. Replace a symmetrical form with an asymmetrical form of the object.
b. If an object is already asymmetrical, increase the degree of asymmetry.
   Example:
   1. One side of a tire is stronger than the other to withstand impact with the curb.
   2. While discharging wet sand through a symmetrical funnel, the sand forms an arch above the opening, causing irregular flow. A funnel of asymmetrical shape completely eliminates the arching effect.

5. Combining
a. Combine in space homogeneous objects or objects destined for contiguous operations.
b. Combine in time homogeneous or contiguous operations.
   Example:
   1. The working element of a rotary excavator has special steam nozzles to defrost and soften the frozen ground in a single step.

6. Universality
a. Have the object perform multiple functions, thereby eliminating the need for some other objects.
   Example:
   1. Sofa converts from a sofa in the day to a bed at night.
   2. Minivan seat adjusts to accommodate seating, sleeping or carrying cargo.

7. Nesting
a. Contain an object inside another, which in turn is placed inside a third object.
b. An object passes through a cavity of another object.

Example:
1. Telescoping antenna
2. Stacking chairs (on top of each other for storage)
3. Mechanical pencil with lead stored inside

8. Counterweight
a. Compensate for the object's weight by joining with another object that has a lifting force.
b. Compensate for the object's weight by providing aerodynamic or hydrodynamic forces.
   Example:
   1. Boat with hydrofoils
   2. Racing cars with rear wing to increase downward pressure

9. Prior counteraction
a. If it is necessary to carry out some action, consider a counteraction in advance.
b. If an object must have a tension, provide anti-tension in advance.
   Example:
   1. Reinforced concrete column or floor
   2. Reinforced shaft – in order to make a shaft stronger, it is made out of several pipes that have been previously twisted to a calculated angle.

10. Prior action
a. Carry out the required action in advance, in full or in part.
b. Arrange objects so that they can go into action without time loss while waiting for the action (and from the most convenient position).
   Example:
   1. Utility knife blade made with a groove allowing the dull part of the blade to be broken off to restore sharpness.
   2. Rubber cement in a bottle is difficult to apply neatly and uniformly. Instead, it is formed into a tape so that the proper amount can be applied more easily.

11. Cushion in advance
a. Compensate for the relatively low reliability of an object by countermeasures taken in advance.
   Example:
   1. To prevent shoplifting, the owner of a store attaches to merchandise a special tag containing a magnetized plate. In order for the customer to carry the merchandise out of the store, the plate is demagnetized by the cashier.

12. Equipotentiality
a. Change the condition of work so that an object need not be raised or lowered.
   Example:
   1. Automobile engine oil is changed by workers in a pit (so that expensive lifting equipment is not needed).

13. Inversion
a. Instead of an action dictated by the specifications of the problem, implement an opposite action.
b. Make a moving part of the object or the outside environment immovable and the nonmoving part movable.
c. Turn the object upside down.
   Example:
   1. Abrasively clean parts by vibrating the parts instead of the abrasive.

14. Spheroidality
a. Replace linear parts or flat surfaces with curved ones and cubical shapes with spherical shapes.
b. Use rollers, balls and spirals.
c. Replace a linear motion with a rotating motion; utilize a centrifugal force.
   Example:
   1. A computer mouse utilizes ball construction to transfer linear, two-axis motion into a vector.

15. Dynamicity
a. Make characteristics of an object or outside environment automatically adjust for optimal performance at each stage of operation.
b. Divide an object into elements able to change position relative to each other.
c. If an object is immovable, make it movable or interchangeable.
   Example:
   1. A flashlight can have a flexible gooseneck between the body and the lamp head.
   2. A transport vessel has a body of cylindrical shape. To reduce the draft of a vessel under full load, the body is comprised of two hinged half-cylindrical parts that can be opened.

16. Partial or overdone action
a. If it is difficult to obtain 100% of a desired effect, achieve somewhat more or less to greatly simplify the problem.
   Example:
   1. A cylinder is painted by dipping it into paint, but it is covered by more paint than is desired. Excess paint is removed by rapidly rotating the cylinder.
   2. To obtain uniform discharge of a metallic powder from a bin, the hopper has a special internal funnel which is continually overfilled to provide nearly constant pressure.

17. Moving to a new dimension
a. Remove problems of moving an object in a line by allowing two-dimensional movement (along a plane). Similarly, problems in moving an object in a plane are removed if the object can be changed to allow three-dimensional movement.
b. Use a multilayer assembly of objects instead of a single layer.
c. Incline the object or turn it "on its side."
d. Project images onto neighboring areas or onto the reverse side of the object.

Example:
1. A greenhouse has a concave reflector on the northern part of the house to improve illumination during the day by reflecting sunlight into that part of the house.

18. Mechanical vibration
a. Set an object into oscillation.
b. If oscillation exists, increase its frequency, even as far as ultrasonic.
c. Use the frequency of resonance.
d. Instead of mechanical vibrators, use piezovibrators.
e. Use ultrasonic vibrations in conjunction with an electromagnetic field.
   Example:
   1. To remove a cast from the body without skin injury, a conventional hand saw is replaced with a vibrating knife.
   2. Vibrate a casting mold while it is being filled to improve flow and structural properties.

19. Periodic action
a. Replace a continuous action with a periodic one or impulse.
b. If an action is already periodic, change its frequency.
c. Use pauses between impulses to provide additional action.
   Example:
   1. An impact wrench loosens corroded nuts using impulses rather than a continuous force.
   2. A warning lamp flashes so that it is even more noticeable than if continuously lit.

20. Continuity of useful action
a. Carry out an action without a break – all parts of an object should be constantly operating at full capacity.
b. Remove an idle and intermediate motion.
   Example:
   1. A drill can have cutting edges which allow for the cutting process in forward and reverse directions.

21. Rushing through

a. Perform harmful or hazardous operations at very high speed.
Example:
1. A cutter for thin-wall plastic tubes prevents tube deformation during cutting by running at a very high speed (cuts before the tube has a chance to deform).

22. Convert harm into benefit

a. Utilize a harmful factor or harmful effect of an environment to obtain a positive effect.
b. Remove a harmful factor by combining it with another harmful factor.
c. Increase the amount of harmful action until it ceases to be harmful.
Example:
1. Sand or gravel freezes solid when transported through cold climates. Overfreezing (using liquid nitrogen) embrittles the ice, which permits pouring.
2. When using high-frequency current to heat metal, only the outer layer is heated. This negative effect is now used for surface heat treating.

23. Feedback

a. Introduce feedback.
b. If feedback already exists, reverse it.
Example:
1. Water pressure from a well is maintained by sensing output pressure and turning on a pump if pressure is low.
2. Ice and water are measured separately but must be combined to an exact total weight. Because it is difficult to precisely dispense the ice, it is measured first. The weight of the ice is fed to the water control, which precisely dispenses the amount of water needed.
3. Noise-canceling devices sample noise signals, phase shift them and feed them back to cancel the effect of the noise source.

24. Mediator

a. Use an intermediary object to transfer or carry out an action.
b. Temporarily connect an object to another one that is easy to remove.
Example:
1. To reduce energy loss when applying current to liquid metal, use cooled electrodes and intermediate liquid metals with a lower melting temperature.

25. Self-service

a. Make the object service itself and carry out supplementary and repair operations.
b. Make use of waste material and energy.
Example:
1. To distribute an abrasive material evenly on the face of crushing rollers and to prevent feeder wear, its surface is made out of the same abrasive material.
2. In an electric welding gun, the rod is advanced by a special device. To simplify the system, the rod is advanced by a solenoid controlled by the welding current.

26. Copying

a. Use a simple or inexpensive copy instead of an object which is complex, expensive, fragile or inconvenient to use.
b. Replace an object or system of objects with an optical copy or image. A scale can be used to reduce or enlarge the image.
c. If visible optical copies are used, replace them with infrared or ultraviolet copies.
Example:
1. The height of tall objects can be determined by measuring their shadows.

27. An inexpensive short-life object instead of an expensive durable one

a. Replace an expensive object by a collection of inexpensive ones, compromising other properties (longevity, for instance).

Example:
1. Disposable diapers
2. A disposable mousetrap consists of a bated plastic tube. A mouse enters the trap through a cone-shaped opening. The angled walls do not allowing the mouse out.

28. Replacement of a mechanical system
a. Replace a mechanical system by an optical, acoustical or odor system.
b. Use an electrical, magnetic or electromagnetic field for interaction with the object.
c. Replace fields.
  1. Stationary fields to moving fields
  2. Fixed fields to those changing in time
  3. Random fields to structured ones
d. Use a field in conjunction with ferromagnetic particles.
Example:
1. To increase a bond of metal coating to a thermoplastic material, the process is carried out inside an electromagnetic field to apply force to the metal.

29. Use a pneumatic or hydraulic construction
a. Replace solid parts of an object with gas or liquid. These parts can use air or water for inflation or use air or hydrostatic cushions.
Example:
1. To increase the draft of an industrial chimney, a spiral pipe with nozzles is installed. When air flows through the nozzles, it creates an air-like wall which reduces drag.
2. For shipping fragile products, air-bubble envelopes or foam-like materials are used.

30. Flexible film or thin membranes
a. Replace customary construction with flexible membrane and thin film.

b. Isolate an object from the outside environment with a thin film or fine membrane.
Example:
1. To prevent the loss of water evaporating from the leaves of plants, polyethylene spray is applied. The polyethylene hardens and plant growth improves because the polyethylene film passes oxygen better than water vapor.

31. Use of porous material
a. Make an object porous or use additional porous elements (inserts, covers, etc.).
b. If an object is already porous, fill the pores in advance with some substance.
Example:
1. To avoid pumping coolant to a machine, some of the parts of the machine are filled with porous material (porous powdered steel) soaked in coolant liquid which evaporates when the machine is working, providing short-term and uniform cooling.

32. Changing the color
a. Change the color of an object or its surroundings.
b. Change the translucency of an object or its surroundings.
c. Use colored additives to observe objects or processes which are difficult to see.
d. If such additives are already used, employ luminescent traces or tracer elements.
Example:
1. A transparent bandage enables a wound to be inspected without the dressing being removed.
2. In steel mills, a water curtain is used to protect workers from overheating. But this curtain only protects from infrared rays, so the bright light from the melted steel can easily get through the curtain. A coloring is added to the water to create a filter effect while remaining transparent.

33. Homogeneity

a. Make objects that interact with a primary object out of the same material or a material that is close to it in behavior.
   Example:
   1. The surface of a feeder for abrasive grain is made of the same material that runs through the feeder, allowing it to have a continuous restoration of the surface without being worn out.

34. Rejecting and regenerating parts

a. After it has completed its function or become useless, reject or modify (e.g., discard, dissolve or evaporate) an element of an object.
b. Restore directly any used up part of an object.
   Example:
   1. Bullet casings are ejected after the gun fires.
   2. Rocket boosters separate after serving their function.

35. Transformation of physical and chemical states of an object

a. Change the aggregate state of an object, the concentration of density, the degree of flexibility or the temperature.
   Example:
   1. In a system for brittle friable materials, the surface of the spiral feedscrew is made from an elastic material with two spiral springs. In order to control the process, the pitch of the screw can be changed remotely.

36. Phase transition

a. Implement an effect developed during the phase transition of a substance; for instance, during the change of volume or during liberation or absorption of heat.
   Example:
   1. To control the expansion of ribbed pipes, they are filled with water and cooled to a freezing temperature.

37. Thermal expansion

a. Use expansion or contraction of a material by heat.
b. Use various materials with different coefficients of heat expansion.
   Example:
   1. To control the opening of roof windows in a greenhouse, bimetallic plates are connected to the windows. With a change of temperature, the plates bend and make the windows open or close.

38. Use strong oxidizers

a. Replace normal air with enriched air.
b. Replace enriched air with oxygen.
c. In oxygen or in air, treat a material with ionizing radiation.
d. Use ionized oxygen.
   Example:
   1. To obtain more heat from a torch, oxygen is fed to the torch instead of atmospheric air.

39. Inert environment

a. Replace the normal environment with an inert one.
b. Carry out a process in a vacuum.
   Example:
   1. To prevent cotton from catching fire in a warehouse, it is treated with inert gas during transport to the storage area.

40. Composite materials

a. Replace a homogeneous material with a composite one.
   Example:
   1. Military aircraft wings are made of composites of plastics and carbon fibers for high strength and low weight.

Inventive Principles Ordered by Frequency of Use

35. Transformation of physical and chemical states of an object

10. Prior action

1. Segmentation

28. Replacement of a mechanical system

2. Extraction

15. Dynamicity

19. Periodic action

18. Mechanical vibration

32. Changing the color

13. Inversion

26. Copying

3. Local quality

27. An inexpensive short-lived object instead of an expensive durable one

29. Use a pneumatic or hydraulic construction

34. Rejecting and regenerating parts

16. Partial or overdone action

40. Composite materials

24. Mediator

17. Moving to a new dimension

6. Universality

14. Spheroidality

22. Convert harm to benefit

39. Inert environment

4. Asymmetry

30. Flexibile film or membranes

37. Thermal expansion

36. Phase transition

25. Self-service

11. Cushion in advance

31. Use of porous materials

38. Use strong oxidizers

8. Counterweight

5. Combining

7. Nesting

21. Rushing through

23. Feedback

12. Equipotentiality

33. Homogeneity

9. Prior counteraction

20. Continuity of useful action

Feature to Change \ Undesired Result (Conflict)	1 Weight of moving object	2 Weight of non-moving object	3 Length of moving object	4 Length of non-moving object	5 Area of moving object	6 Area of non-moving object	7 Volume of moving object	8 Volume of non-moving object	9 Speed	10 Force	11 Tension, pressure	12 Shape	13 Stability of object
1 Weight of moving object			15,8, 29,34		29,17, 38,34		29,2, 40,28		2,8, 15,38	8,10, 18,37	10,36, 37,40	10,14, 35,40	1,35, 19,39
2 Weight of non-moving object				10,1, 29,35		35,30, 13,2		5,35, 14,2		8,10, 19,35	13,29, 10,18	13,10, 29,14	26,39, 1,40
3 Length of moving object	8,15, 29,34				15,17, 4		7,17, 4,35		13,4, 8	17,10, 4	1,8, 35	1,8, 10,29	1,8, 15,34
4 Length of non-moving object		35,28, 40,29			17,7, 10,40		35,8, 2,14			28,10	1,14, 35	13,14, 15,7	39,37, 35
5 Area of moving object	2,17, 29,4		14,15, 18,4				7,14, 17,4		29,30, 4,34	19,30, 35,2	10,15, 36,28	5,34, 29,4	11,2, 13,39
6 Area of non-moving object		30,2, 14,18		26,7, 9,39						1,18, 36,37	10,15, 36,37		2,38
7 Volume of moving object	2,26, 29,40		1,7, 4,35		1,7, 4,17				29,4, 38,34	15,35, 36,37	6,35, 36,37	1,15, 29,4	28,10, 1,39
8 Volume of non-moving object		35,10, 19,14	19,14	35,8, 2,14						2,18, 37	24,35	7,2, 35	34,28, 35,40
9 Speed	2,28, 13,38		13,14, 8		29,30, 34		7,29, 34			13,28, 15,19	6,18, 38,40	35,15, 18,34	28,33, 1,18
10 Force	8,1, 37,18	18,13, 1,28	17,19, 9,36	28,10	19,10, 15	1,18, 36,37	15,9, 12,37	2,36, 18,37	13,28, 15,12		18,21, 11	10,35, 40,34	35,10, 21
11 Tension, pressure	10,36, 37,40	13,29, 10,18	35,10, 36	35,1, 14,16	10,15, 36,28	10,15, 36,37	6,35, 10	35,24	6,35, 36	36,35, 21		35,4, 15,10	35,33, 2,40
12 Shape	8,10, 29,40	15,10, 26,3	29,34, 5,4	13,14, 10,7	5,34, 4,10		14,4, 15,22	7,2, 35	35,15, 34,18	35,10, 37,40	34,15, 10,14		33,1, 18,4
13 Stability of object	21,35, 2,39	26,39, 1,40	13,15, 1,28	37	2,11, 13	39	28,10, 19,39	34,28, 35,40	33,15, 28,18	10,35, 21,16	2,35, 40	22,1, 18,4	
14 Strength	1,8, 40,15	40,26, 27,1	1,15, 8,35	15,14, 28,26	3,34, 40,29	9,40, 28	10,15, 14,7	9,14, 17,15	8,13, 26,14	10,18, 3,14	10,3, 18,40	10,30, 35,40	13,17, 35
15 Durability of moving object	19,5, 34,31		2,19, 9		3,17, 19		10,2, 19,30		3,35, 5	19,2, 16	19,3, 27	14,26, 28,25	13,3, 35
16 Durability of non-moving object		6,27, 19,16		1,10, 35				35,34, 38					39,3, 35,23
17 Temperature	36,22, 6,38	22,35, 32	15,19, 9	15,19, 9	3,35, 39,18	35,38	34,39, 40,18	35,6, 4	2,28, 36,30	35,10, 3,21	35,39, 19,2	14,22, 19,32	1,35, 32
18 Brightness	19,1, 32	2,35, 32	19,32, 16		19,32, 26		2,13, 10		10,13, 19	26,19, 6		32,30	32,3, 27
19 Energy spent by moving object	12,18, 28,31		12,28		15,19, 25		35,13, 18		8,15, 35	16,26, 21,2	23,14, 25	12,2, 29	19,13, 17,24
20 Energy spent by non-moving object		19,9, 6,27								36,37			27,4, 29,18

40 TRIZ Principles and Contradiction Table

Undesired Result (Conflict) / Feature to Change	14 Strength	15 Durability of moving object	16 Durability of non-moving object	17 Temperature	18 Brightness	19 Energy spent by moving object	20 Energy spent by non-moving object	21 Power	22 Waste of energy	23 Waste of substance	24 Loss of information	25 Waste of time	26 Amount of substance
1 Weight of moving object	28,27,18,40	5,34,31,35		6,20,4,38	19,1,32	35,12,34,31		12,36,18,31	6,2,34,19	5,35,3,31	10,24,35	10,35,20,28	3,26,18,31
2 Weight of non-moving object	28,2,10,27		2,27,19,6	28,19,32,22	19,32,35		18,19,28,1	15,19,18,22	18,19,28,15	5,8,13,30	10,15,35	10,20,35,26	19,6,18,26
3 Length of moving object	8,35,29,34	19		10,15,19	32	8,35,24		1,35	7,2,35,39	4,29,23,10	1,24	15,2,29	29,35
4 Length of non-moving object	15,14,28,26		1,40,35	3,35,38,18	3,25			12,8	6,28	10,28,24,35	24,26	30,29,14	
5 Area of moving object	3,15,40,14	6,3		2,15,16	15,32,19,13	19,32		19,10,32,18	15,17,30,26	10,35,2,39	30,26	26,4	29,30,6,13
6 Area of non-moving object	40		2,10,19,30	35,39,38				17,32	17,7,30	10,14,18,39	30,16	10,35,4,18	2,18,40,4
7 Volume of moving object	9,14,15,7	6,35,4		34,39,10,18	2,13,10	35		35,6,13,18	7,15,13,16	36,39,34,10	2,22	2,6,34,10	29,30,7
8 Volume of non-moving object	9,14,17,15		35,34,38	35,6,4				30,6		10,39,35,34		35,16,32,18	35,3
9 Speed	8,3,26,14	3,19,35,5		28,30,36,2	10,13,19	8,15,35,38		19,35,38,2	14,20,19,35	10,13,28,38	13,26		18,19,29,38
10 Force	35,10,14,27	19,2		35,10,21		19,17,10	1,16,36,37	19,35,18,37	14,15	8,35,40,5		10,37,36	14,29,18,36
11 Tension, pressure	9,18,3,40	19,3,27		35,39,19,2		14,24,10,37		10,35,14	2,36,25	10,36,3,37		37,36,4	10,14,36
12 Shape	30,14,10,40	14,26,9,25		22,14,19,32	13,15,32	2,6,34,14		4,6,2	14	35,29,3,5		14,10,34,17	36,22
13 Stability of object	17,9,15	13,27,10,35	39,3,35,23	35,1,32	32,3,27,15	13,19	27,4,29,18	32,35,27,31	14,2,39,6	2,14,30,40		35,27	15,32,35
14 Strength		27,3,26		30,10,40	35,19	19,35,10	35	10,26,35,28	35	35,28,31,40		29,3,28,10	29,10,27
15 Durability of moving object	27,3,10			19,35,39	2,19,4,35	28,6,35,18		19,10,35,38		28,27,3,18	10	20,10,28,18	3,35,10,40
16 Durability of non-moving object				19,18,36,40			16	16		27,16,18,38	10	28,20,10,16	3,35,31
17 Temperature	10,30,22,40	19,13,39	19,18,36,40		32,30,21,16	19,15,3,17		2,14,17,25	21,17,35,38	21,36,29,31		35,28,21,18	3,17,30,39
18 Brightness	35,19	2,19,6		32,35,19		32,1,19	32,35,1,15	32	19,16,1,6	13,1	1,6	19,1,26,17	1,19
19 Energy spent by moving object	5,19,9,35	28,35,6,18		19,24,3,14	2,15,19			6,19,37,18	12,22,15,24	35,24,18,5		35,38,19,18	34,23,16,18
20 Energy spent by non-moving object	35				19,2,35,32					28,27,18,31			3,35,31

188

#	Feature to Change \ Undesired Result (Conflict)	27 Reliability	28 Accuracy of measurement	29 Accuracy of manufacturing	30 Harmful factors acting on object	31 Harmful side effects	32 Manufacturability	33 Convenience of use	34 Repairability	35 Adaptability	36 Complexity of device	37 Complexity of control	38 Level of automation	39 Productivity
1	Weight of moving object	3,11, 1,27	28,27, 35,26	28,35, 26,18	22,21, 18,27	22,35, 31,39	27,28, 1,36	35,3, 2,24	2,27, 28,11	29,5, 15,8	26,30, 36,34	28,29, 26,32	26,35, 18,19	35,3, 24,37
2	Weight of non-moving object	10,28, 8,3	18,26, 28	10,1, 35,17	2,19, 22,37	35,22, 1,39	28,1, 9	6,13, 1,32	2,27, 28,11	19,15, 29	1,10, 26,39	25,28, 17,15	2,26, 35	1,28, 15,35
3	Length of moving object	10,14, 29,40	28,32, 4	10,28, 29,37	1,15, 17,24	17,15	1,29, 17	15,29, 35,4	1,28, 10	14,15, 1,16	1,19, 26,24	35,1, 26,24	17,24, 26,16	14,4, 28,29
4	Length of non-moving object	15,29, 28	32,28, 3	2,32, 10	1,18		15,17, 27	2,25	3	1,35	1,26	26		30,14, 7,26
5	Area of moving object	29,9	26,28, 32,3	2,32	22,33, 28,1	17,2, 18,39	13,1, 26,24	15,17, 13,16	15,13, 10,1	15,30	14,1, 13	2,36, 26,18	14,30, 28,23	10,26, 34,2
6	Area of non-moving object	32,35, 40,4	26,28, 32,3	2,29, 18,36	27,2, 39,35	22,1, 40	40,16	16,4	16	15,16	1,18, 36	2,35, 30,18	23	10,15, 17,7
7	Volume of moving object	14,1, 40,11	25,26, 28	25,28, 2,16	22,21, 27,35	17,2, 40,1	29,1, 40	15,13, 30,12	10	15,29	26,1	29,26, 4	35,34, 16,24	10,6, 2,34
8	Volume of non-moving object	2,35, 16		35,10, 25	34,39, 19,27	30,18, 35,4	35		1		1,31	2,17, 26		35,37, 10,2
9	Speed	11,35, 27,28	28,32, 1,24	10,28, 32,25	1,28, 35,23	2,24, 35,21	35,13, 8,1	32,28, 13,12	34,2, 28,27	15,10, 26	10,28, 4,34	3,34, 27,16	10,18	
10	Force	3,35, 13,21	35,10, 23,24	28,29, 37,36	1,35, 40,18	13,3, 36,24	15,37, 18,1	1,28, 3,25	15,1, 11	15,17, 18,20	26,35, 10,18	36,37, 10,19	2,35	3,28, 35,37
11	Tension, pressure	10,13, 19,35	6,28, 25	3,35	22,2, 37	2,33, 27,18	1,35, 16	11	2	35	19,1, 35	2,36, 37	35,24	10,14, 35,37
12	Shape	10,40, 16	28,32, 1	32,30, 40	22,1, 2,35	35,1	1,32, 17,28	32,15, 26	2,13, 1	1,15, 29	16,29, 1,28	15,13, 39	15,1, 32	17,26, 34,10
13	Stability of object		13	18	35,24, 30,18	35,40, 27,39	35,19	32,35, 30	2,35, 10,16	35,30, 34,2	2,35, 22,26	35,22, 39,23	1,8, 35	23,35, 40,3
14	Strength	11,3	3,27, 16	3,27	18,35, 37,1	15,35, 22,2	11,3, 10,32	32,40, 28,2	27,11, 3	15,3, 32	2,13, 28	27,3, 15,40	15	29,35, 10,14
15	Durability of moving object	11,2, 13	3	3,27, 16,40	22,15, 33,28	21,39, 16,22	27,1, 4	12,27	29,10, 27	1,35, 13	10,4, 29,15	19,29, 39,35	6,10	35,17, 14,19
16	Durability of non-moving object	34,27, 6,40	10,26, 24		17,1, 40,33	22	35,10	1	1	2		25,34, 6,35	1	20,10, 16,38
17	Temperature	19,35, 3,10	32,19, 24	24	22,33, 35,2	22,35, 2,24	26,27	26,27	4,10, 16	2,18, 27	2,17, 16	3,27, 35,31	26,2, 19,16	15,28, 35
18	Brightness		11,15, 32	3,32	15,19	35,19, 32,39	19,35, 28,26	28,26, 19	15,17, 13,16	15,1, 19	6,32, 13	32,15	2,26, 10	2,25, 16
19	Energy spent by moving object	19,21, 11,27	3,1, 32		1,35, 6,27	2,35, 6	28,26, 30	19,35	1,15, 17,28	15,17, 13,16	2,29, 27,28	35,38	32,2	12,28, 35
20	Energy spent by non-moving object	10,36, 23			10,2, 22,37	19,22, 18	1,4					19,35, 16,25		1,6

Feature to Change	1 Weight of moving object	2 Weight of non-moving object	3 Length of moving object	4 Length of non-moving object	5 Area of moving object	6 Area of non-moving object	7 Volume of moving object	8 Volume of non-moving object	9 Speed	10 Force	11 Tension, pressure	12 Shape	13 Stability of object
21 Power	8,36, 38,31	19,26, 17,27	1,10, 35,37		19,38	17,32, 13,38	35,6, 38	30,6, 25	15,35, 2	26,2, 36,35	22,10, 35	29,14, 2,40	35,32, 15,31
22 Waste of energy	15,6, 19,28	19,6, 18,9	7,2, 6,13	6,38, 7	15,26, 17,30	17,7, 30,18	7,18, 23	7	16,35, 38	36,38			14,2, 39,6
23 Waste of substance	35,6, 23,40	35,6, 22,32	14,29, 10,39	10,28, 24	35,2, 10,31	10,18, 39,31	1,29, 30,36	3,39, 18,31	10,13, 28,38	14,15, 18,40	3,36, 37,10	29,35, 3,5	2,14, 30,40
24 Loss of information	10,24, 35	10,35, 5	1,26	26	30,26	30,16		2,22	26,32				
25 Waste of time	10,20, 37,35	10,20, 26,5	15,2, 29	30,24, 14,5	26,4, 5,16	10,35, 17,4	2,5, 34,10	35,16, 32,18		10,37, 36,5	37,36, 4	4,10, 34,17	35,3, 22,5
26 Amount of substance	35,6, 18,31	27,26, 18,35	29,14, 35,18		15,14, 29	2,18, 40,4	15,20, 29		35,29, 34,28	35,14, 3	10,36, 14,3	35,14	15,2, 17,40
27 Reliability	3,8, 10,40	3,10, 8,28	15,9, 14,4	15,29, 28,11	17,10, 14,16	32,35, 40,4	3,10, 14,24	2,35, 24	21,35, 11,28	8,28, 10,3	10,24, 35,19	35,1, 16,11	
28 Accuracy of measurement	32,35, 26,28	28,35, 25,26	28,26, 5,16	32,28, 3,16	26,28, 32,3	26,28, 32,3	32,13, 6		28,13, 32,24	32,2	6,28, 32	6,28, 32	32,35, 13
29 Accuracy of manufacturing	28,32, 13,18	28,35, 27,9	10,28, 29,37	2,32, 10	28,33, 29,32	2,29, 18,36	32,28, 2	25,10, 35	10,28, 32	28,19, 34,36	3,35	32,30, 40	30,18
30 Harmful factors acting on object	22,21, 27,39	2,22, 13,24	17,1, 39,4	1,18	22,1, 33,28	27,2, 39,35	22,23, 37,35	34,39, 19,27	21,22, 35,28	13,35, 39,18	22,2, 37	22,1, 3,35	35,24, 30,18
31 Harmful side effects	19,22, 15,39	35,22, 1,39	17,15, 16,22		17,2, 18,39	22,1, 40	17,2, 40	30,18, 35,4	35,28, 3,23	35,28, 1,40	2,33, 27,18	35,1	35,40, 27,39
32 Manufacturability	28,29, 15,16	1,27, 36,13	1,29, 13,17	15,17, 27	13,1, 26,12	16,40	13,29, 1,40	35	35,13, 8,1	35,12	35,19, 1,37	1,28, 13,27	11,13, 1
33 Convenience of use	25,2, 13,15	6,13, 1,25	1,17, 13,12		1,17, 13,16	18,16, 15,39	1,16, 35,15	4,18, 39,31	18,13, 34	28,13, 35	2,32, 12	15,34, 29,28	32,35, 30
34 Repairability	2,27, 35,11	2,27, 35,11	1,28, 10,25	3,18, 31	15,13, 32	16,25	25,2, 35,11	1	34,9	1,11, 10	13	1,13, 2,4	2,35
35 Adaptability	1,6, 15,8	19,15, 29,16	35,1, 29,2	1,35, 16	35,30, 29,7	15,16	15,35, 29	35	35,10, 14	15,17, 20	35,16	15,37, 1,8	35,30, 14
36 Complexity of device	26,30, 34,36	2,36, 35,39	1,19, 26,24	26	14,1, 13,16	6,36	34,26, 6	1,16	34,10, 28	26,16	19,1, 35	29,13, 28,15	2,22, 17,19
37 Complexity of control	27,26, 28,13	6,13, 28,1	16,17, 26,24	26	2,13, 15,17	2,39, 30,16	29,1, 4,16	2,18, 26,31	3,4, 16,35	36,28, 40,19	35,36, 37,32	27,13, 1,39	11,22, 39,30
38 Level of automation	28,26, 18,35	28,26, 35,10	14,13, 17,28	23	17,14, 13		35,13, 16		28,10	2,35	13,35	15,32, 1,13	18,1
39 Productivity	35,26, 24,37	28,27, 15,3	18,4, 28,38	30,7, 14,26	10,26, 34,31	10,35, 17,7	2,6, 34,10	35,37, 10,2		28,15, 10,36	10,37, 14	14,10, 34,40	35,3, 22,39

Feature to Change \ Undesired Result (Conflict)	14 Strength	15 Durability of moving object	16 Durability of non-moving object	17 Temperature	18 Brightness	19 Energy spent by moving object	20 Energy spent by non-moving object	21 Power	22 Waste of energy	23 Waste of substance	24 Loss of information	25 Waste of time	26 Amount of substance
21 Power	26,10,28	19,35,10,38	16	2,14,17,25	16,6,19	16,6,19,37			10,35,38	28,27,18,38	10,19	35,20,10,6	4,34,19
22 Waste of energy	26			19,38,7	1,13,32,15			3,38		35,27,2,37	19,10	10,18,32,7	7,18,25
23 Waste of substance	35,28,31,40	28,27,3,18	27,16,18,38	21,36,39,31	1,6,13	35,18,24,5	28,27,12,31	28,27,18,38	35,27,2,31			15,18,35,10	6,3,10,24
24 Loss of information		10	10		19		1	10,19	19,10			24,26,28,32	24,28,35
25 Waste of time	29,3,28,18	20,10,28,18	28,20,10,16	35,29,21,18	1,19,26,17	35,38,19,18	1	35,20,10,6	10,5,18,32	35,18,10,39	24,26,28,32		35,38,18,16
26 Amount of substance	14,35,34,10	3,35,10,40	3,35,31	3,17,39		34,29,16,18	3,35,31	35	7,18,25	6,3,10,24	24,28,35	35,38,18,16	
27 Reliability	11,28	2,35,3,25	34,27,6,40	3,35,10	11,32,13	21,11,27,19	36,23	21,11,26,31	10,11,35	10,35,29,39	10,28	10,30,4	21,28,40,3
28 Accuracy of measurement	28,6,32	28,6,32	10,26,24	6,19,28,24	6,1,32	3,6,32		3,6,32	26,32,27	10,16,31,28		24,34,28,32	2,6,32
29 Accuracy of manufacturing	3,27	3,27,40		19,26	3,32	32,2		32,2	13,32,2	35,31,10,24		32,26,28,18	32,30
30 Harmful factors acting on object	18,35,37,1	22,15,33,28	17,1,40,33	22,33,35,2	1,19,32,13	1,24,6,27	10,2,22,37	19,22,31,2	21,22,35,2	33,22,19,40	22,10,2	35,18,34	35,33,29,31
31 Harmful side effects	15,35,22,2	15,22,33,31	21,39,16,22	22,35,2,24	19,24,39,32	2,35,6	19,22,18	2,35,18	21,35,2,22	10,1,34	10,21,29	1,22	3,24,39,1
32 Manufacturability	1,3,10,32	27,1,4	35,16	27,26,18	28,24,27,1	28,26,27,1	1,4	27,1,12,24	19,35	15,34,33	32,24,18,16	35,28,34,4	35,23,1,24
33 Convenience of use	32,40,3,28	29,3,8,25	1,16,25	26,27,13	13,17,1,24	1,13,24		35,34,2,10	2,19,13	28,32,2,24	4,10,27,22	4,28,10,34	12,35
34 Repairability	11,1,2,9	11,29,28,27	1	4,10	15,1,13	15,1,28,16		15,10,32,2	15,1,32,19	2,35,34,27		32,1,10,25	2,28,10,25
35 Adaptability	35,3,32,6	13,1,35	2,16	27,2,3,35	6,22,26,1	19,35,29,13		19,1,29	18,15,1	15,10,2,13	35,33	35,28	3,35,15
36 Complexity of device	2,13,28	10,4,28,15		2,17,13	24,17,13	27,2,29,28		20,19,30,34	10,35,13,2	35,10,28,29	35,33	6,29	13,3,27,10
37 Complexity of control	27,3,15,28	19,29,39,25	25,24,6,35	3,27,35,16	2,24,26	35,38	19,35,16	19,1,16,10	35,3,15,19	1,13,10,24	35,33,27,22	18,28,32,9	3,27,29,18
38 Level of automation	25,13	6,9		26,2,19	8,32,19	2,32,13		28,2,27	23,28	35,10,18,5	35,33	24,28,35,30	35,13
39 Productivity	29,28,10,18	35,10,2,18	20,10,16,38	35,21,28,10	26,17,19,1	35,10,38,19	1	35,20,10	28,10,29,35	28,10,35,23	13,15,23		35,38

Feature to Change \ Undesired Result (Conflict)	27 Reliability	28 Accuracy of measurement	29 Accuracy of manufacturing	30 Harmful factors acting on object	31 Harmful side effects	32 Manufacturability	33 Convenience of use	34 Repairability	35 Adaptability	36 Complexity of device	37 Complexity of control	38 Level of automation	39 Productivity
21 Power	19,24,26,31	32,15,2	32,2	19,22,31,2	2,35,18	26,10,34	26,35,10	35,2,10,34	19,17,34	20,19,30,34	19,35,16	28,2,17	28,35,34
22 Waste of energy	11,10,35	32		21,22,35,2	21,35,2,22		35,32,1	2,19		7,23	35,3,15,23	2	28,10,29,35
23 Waste of substance	10,29,39,35	16,34,31,28	35,10,24,31	33,22,30,40	10,1,34,29	15,34,33	32,28,2,24	2,35,34,27	15,10,2	35,10,28,24	35,18,10,13	35,10,18	28,35,10,23
24 Loss of information	10,28,23			22,10,1	10,21,22	32	27,22				35,33	35	13,23,15
25 Waste of time	10,30,4	24,34,28,32	24,26,28,18	35,18,34	35,22,18,39	35,28,34,4	4,28,10,34	32,1,10	35,28	6,29	18,28,32,10	24,28,35,30	
26 Amount of substance	18,3,28,40	13,2,28	33,30	35,33,29,31	3,35,40,39	29,1,35,27	35,29,25,10	2,32,10,25	15,3,29	3,13,27,10	3,27,29,18	8,35	13,29,3,27
27 Reliability		32,3,11,23	11,32,1	27,35,2,40	35,2,40,26		27,17,40	1,11	13,35,8,24	13,35,1	27,40,28	11,13,27	1,35,29,38
28 Accuracy of measurement	5,11,1,23			28,24,22,26	3,33,39,10	6,35,25,18	1,13,17,34	1,32,13,11	13,35,2	27,35,10,34	26,24,32,28	28,2,10,34	10,34,28,32
29 Accuracy of manufacturing	11,32,1	26,28,10,36		26,28,10,18	4,17,34,26		1,32,35,23	25,10		26,2,18		26,28,18,23	10,18,32,39
30 Harmful factors acting on object	27,24,2,40	28,33,23,26	26,28,10,18		24,35,2		2,25,28,39	35,10,2	35,11,22,31	22,19,29,40	22,19,29,40	33,3,34	22,35,13,24
31 Harmful side effects	24,2,40,39	3,33,26	4,17,34,26							19,1,31	2,21,27,1	2	22,35,18,39
32 Manufacturability	1,35,12,18			24,2			2,5,13,16	35,1,11,9	2,13,15	27,26,1	6,28,11,1	8,28,1	35,1,10,28
33 Convenience of use	17,27,8,40	25,13,2,34	1,32,35,23	2,25,28,39		2,5,12		12,26,1,32	15,34,1,16	32,26,12,17		1,34,12,3	15,1,28
34 Repairability	11,10,1,16	10,2,13	25,10	35,10,2,16		1,35,11,10	1,12,26,15		7,1,4,16	35,1,13,11		34,35,7,13	1,32,10
35 Adaptability	35,13,8,24	35,5,1,10	35,11,22,31			1,13,31	15,34,1,16	1,16,7,4		15,29,37,28	1	27,34,35	35,28,6,37
36 Complexity of device	13,35,1	27,35,10,34	26,24,32	22,19,29,40	19,1	27,26,1,13	27,9,26,24	1,13	29,15,28,37		15,10,37,28	15,1,24	12,17,28
37 Complexity of control	27,40,28,8	26,24,32,28		22,19,29,28	2,21	5,28,11,29	2,5	12,26	1,15	15,10,37,28		34,21	35,18
38 Level of automation	11,27,32	28,26,10,34	28,26,18,23	2,33	2	1,26,13	1,12,34,3	1,35,13	27,4,1,35	15,24,10	34,27,25		5,12,35,26
39 Productivity	1,35,10,38	1,10,34,28	18,10,32,1	22,35,13,24	35,22,18,39	35,28,2,24	1,28,7,19	1,32,10,25	1,35,28,37	12,17,28,24	35,18,27,2	5,12,35,26	

Appendix C

E-Mail Addresses

Join QFD list server
qfd-l@quality.org
by sending the message
subscribe QFD-L to
majordomo@quality.org

e-mail for QFD Institute
qfdi@qfdi.org

e-mail for John Terninko
john@terninko.com

web page for John Terninko
http://www.mv/ipusers/rm

GLOSSARY

Acceptance matrix: a matrix for the manufacturing process which contains cells representing a manufacturing product test's ability to predict a product's performance for a particular performance measure.

Adjustment factor: a component of Dr. Taguchi's Dynamic Robust Design that modifies a design's sensitivity to the signal factor (see signal factor).

Affinity diagram: provides a structure for organizing data into related groups.

American Supplier Institute (ASI): one of two nonprofit US quality organizations. The American Supplier Institute was formed in 1983, with Larry Sullivan as chairperson. ASI started with statistics and added Taguchi's Quality Engineering philosophy.

Analog/digital: an analog presentation is graphic and pictorial, as opposed to a digital presentation which is numeric.

Analogic thinking: the mental process of drawing on and applying the solution of a known problem to one currently faced.

Analytical Hierarchy Process (AHP): a formal procedure for ranking alternatives so that the scores are of a variable nature.

Anticipatory Failure Determination: a method for systematically identifying and eliminating system failures – before they occur. This method in effect invents failure mechanisms and then examines the possibility of their actually occurring.

Benchmarking: a formal analysis comparing one organization's product to the competition. The result is the identification of the best industrial performance.

Brainstorm: a process used to generate as many ideas as possible.

CAPD (Check, Act, Plan Do): a reactive control cycle that analyzes problems after they have occurred.

Central Japan Quality Organization: a quality organization that trains some Toyota suppliers in quality control.

Concept selection: the process of deciding which alternative design of a system is most appropriate based upon certain criteria.

Consumer: a person or organization that purchases and uses a product or service.

Contradiction: The condition wherein contradictory requirements are placed upon a technological system. Others give the name paradox or oxymoron to such situations.

Correlation: when changes in one factor seem to have a predictive relationship with another factor (i.e., the taller a person, the greater his weight).

Countermeasures: proactive modifications to a design or a process that reduce the likelihood of a particular failure reoccurring.

Customer: the individual or organization that purchases or influences the purchase of a product or service.

Customer's verbatim response: the actual words a customer uses when evaluating or commenting on a product or service.

Demanded quality: the customer's subjective description of the performance of the product and its functions. Grammatically, it contains an adverb and verb. Demanded qualities should always be worded as positive statements for ease in subsequent analysis.

Design of experiments: a systematic way to evaluate combinations of factors to develop a mathematical model for the performance of a response.

Design FMEA: an FMEA done on the parameter values of the design.

Directed Product Evolution: the systematic application of the Patterns of Evolution of Technological Systems to a current system to "force" its highly probable future development – before it occurs naturally. In effect, we systematically invent the future. Via this process, dominant market and patent positions can be systematically achieved.

Directed Product Improvement: improve systems by moving their designs in the direction of ideality.

DQ: demanded quality.

Dynamic Robust Design: Dr. Taguchi's approach enabling the production of a design that can be easily modified.

Failure mode: a type of defect in a product. Failure modes are often expressed as negative statements.

Fault tree: a structured way of showing relationships between failures and causes.

FMEA (Failure Mode and Effect Analysis): prioritizes manufacturing or design causes of failure. Corrective action is taken for the more important causes.

Focus group: a group of customers that uses and evaluates a product and the competitors' products in front of a team. These groups often help QFD practitioners observe previously overlooked design flaws.

Function: what a product does or what task is performed. Grammatically, a function has an active verb plus an object.

Functions of an organization: the different organizational units specializing in some field, such as marketing or engineering.

Gemba: the customer's/user's environment; the context in which the product is used.

GOAL/QPC: Growth Opportunity Alliance of Lawrence/Quality, Productivity, Competitiveness. One of three nonprofit quality organizations in the US. GOAL/QPC is a center for seminars and training. GOAL/QPC was established in 1978 to promote quality in the industries of Lawrence, Massachusetts.

Grand average: average of all the averages.

Hin Shitsu Ki No Ten Kai: the six Japanese kanji characters translated to mean Quality Function Deployment.

Hoshin Kanri: a corporate vision that drives vertical deployment to attain a few key quality improvements for the organization.

House of Quality: a matrix incorporating the customer's demanded qualities and the product's performance measures. This manual refers to the House of Quality as the Demanded Qualities vs. Performance Measures Matrix. This matrix enables QFD practitioners to consult only one diagram in order to scrutinize the relationships between the customer's demands and the design team's efforts to meet them.

Ideality: a qualitative assessment defined as the sum of a system's useful functions divided by the sum of its undesired characteristics (drawbacks). A perfect system provides all desired characteristics with no drawbacks. The use of resources and physical, chemical, geometrical and other effects makes the ideal possible.

Independence (of two or more factors): occurs when the effect of one factor is not influenced by the other factor.

Interaction (of two or more factors): occurs when the effect of one factor is influenced by the existence of another factor.

Interval numbers: a measurement system that indicates the order in which the difference between numbers has meaning. Interval numbers can be used for counting frequency. Calculating the mean (average), medium and mode are also appropriate.

Inventive problem: a problem which includes a contradiction and for which a path to a solution is unknown.

Kano Model: a tool that determines the importance of demanded qualities by sorting them into three categories: basic needs, performance needs and excitement needs.

Kansei: sensory engineering used to create product designs that appeal to the customer's sense of smell, touch, etc. Kansei often involves some intangible attributes, such as communicating a feeling of luxury or a particular image.

KJ: refers to a comprehensive process used to analyze textual information. Often mistakenly believed to be the same as the affinity diagram. The process was developed by Jiro Kawakita, a cultural anthropologist.

Left brain: generally refers to logical and linear thought processes.

Level of innovation: TRIZ partitions problem solutions into five levels: Standard solutions are termed Level 1. Improvements – Level 2. Innovations – Level 3. Inventions – Level 4. Discoveries – Level 5.

Loss function: a way of measuring quality developed by Taguchi that establishes a financial measure of the customer's dissatisfaction with a product's performances as it deviates from a target value. The function includes both the variance and the deviation of the average performance from the target value.

Manufacturing FMEA: an FMEA that looks at the manufacturing process.

Matrix: a pictorial way of looking at the relationships between two or more factors. In most applications it has columns and rows with all the intersections included.

Mean, arithmetic: the average.

Mean, geometric: the nth root of the product of n numbers.

Measurement system: generic term for the collection of numeric methodologies with differing information values that enable different calculations.

Noise factor: Dr. Taguchi's term for uncontrolled sources of variation that affect the manufacturing process.

Nominal numbers: a measurement system used to identify items; counting the frequency of each category is possible with nominal numbers.

Normalized score: a percent or fraction of the total of all scores; same as relative frequency.

Open communication: free flow of information, ideas and feelings without fear of punishment.

Glossary

Ordinal numbers: a measurement system used to identify and designate order for a collection of items. Counting frequency of categories and calculating the median and mode are possible. Finding the average is not appropriate.

Orthogonal array: a balanced set of combinations used to select factor levels for conducting an efficient investigation. It could be used in a designed experiment.

Patterns of Evolution of Technological Systems: a compilation of trends within TRIZ which document historically recurring strong tendencies in the development of man-made systems. These patterns can be applied to virtually any design to understand its probable future and to accelerate that future's realization.

PDCA (Plan, Do, Check, Act): a proactive control cycle that is a fundamental QFD building block. This tool helps teams avoid potential problems before they occur.

Performance measure: technical measurement evaluating the performance of some demanded quality.

Physical contradiction: a contradiction wherein some element of a system has two opposing requirements.

PM: performance measure

Predictive value: the results expected from a second factor when another factor is known.

Product: a tangible item.

Product Design Process Chart: contains the critical processes and the critical tasks of the design process. Columns present the organization's functional units, rectangles identify who is responsible for the various activities and arrows indicate the flow of documents or decisions.

Product Planning Table: one of the components of the Demanded Qualities vs. Performance Measures Matrix. The table calculates the relative importance of the performance measures and sets target values.

Pugh's concept selection: a method for evaluating alternative designs and generating new designs.

Quality circles: groups formed to clarify and correct problems on the work floor. Group members voluntarily choose the problems, activities or the means that they pursue. Quality circles are based on the idea of autonomous control, with the goal of improving the work area and productivity.

Quality Planning Table: one of the components of the Demanded Qualities vs Performance Measures Matrix. The table compares an organization's product against the competition in order to establish the composite importance for each demanded quality.

QFD Institute: a North American quality organization, the QFD Institute emerged in 1993 as an outgrowth of experiments conducted by John Terninko, Richard Zultner, Glenn Mazur and Satoshi Nakui. The institute created the first North American QFD Master Class, held in Dearborn, Michigan in 1994.

Ratio numbers: the ratio of two numbers has meaning because there is an absolute zero, and the interval between numbers is meaningful.

Relationship Matrix: a component of the Demanded Qualities vs. Performance Measures Matrix that identifies the strength of each performance measure's ability to predict the customer's satisfaction with a particular demanded quality.

Relative frequency: ratio of the number of occurrences of some event to the total number of occurrences of all events.

Resources: those system/environment elements and their attributes, and sources of energy (heat, electricity, magnetism, motion, etc.) within and around (available to) the system, that can be used in moving the sytem toward ideality (i.e., its improvement).

Right brain: generally refers to creative and nonlinear thought processes.

Robust Design: in the context of Taguchi's Quality Engineering, a robust design has reduced sensitivity to uncontrolled factors or variations. The Japanese term for robust design is quality engineering.

Sales points: specific features that will distinguish a product from the competition.

SDCA (Standardize, Do, Check, Act): a fundamental QFD building block; a proactive cycle that helps teams avoid potential problems.

Service: an intangible item.

Signal factor: a component of Dr. Taguchi's Dynamic Robust Design that moves the product's performance in a given area to some particular value.

Solution: a specific design, technology, methodology, manufacturing process or material to be used.

Specification: a level of performance for a component, subsystem or system necessary for satisfactory performance of the product.

Stakeholder: anyone who can influence the decision to use or buy the product and anyone who is impacted by the use of the product.

Standard Solutions: TRIZ knowledge base that has been developed which includes a compilation of contradictions and their solutions, at an abstract level, that have been repetitively and successfully applied in the past, as well as typical resources used in their implementation.

Static Robust Design: Dr. Taguchi's tool enabling a product design or manufacturing process to attain one fixed performance level; see Dynamic Robust Design (see also adjustment factor).

Strength of relationship: a subjective measure of the predictive ability of one factor on another factor. Informally, this is similar to the correlation coefficient.

Subjective benchmarking: using the consumer's/user's opinion to determine the industrial best.

Substitute quality characteristics: an indirect but objective way to measure a product's performance in a given area when a direct test is not possible. For instance, putting a soft plastic in an oven is supposed to simulate time for the migration of plasticizers.

Target value: a desired performance level for directing development. It may be better or worse than the final selected specification.

Technical benchmarking: measures of performance using measurable, variable data to compare different producers of the same product.

Glossary

Technical contradictions: contradictions wherein an improvement in one desired characteristic of a system results in the deterioration of another.

Tree: a diagram used to ensure that the items in a group of data are at the same level of abstraction and that there are no missing data. The tree is often used with the affinity diagram. The sequence can go from the branches to the trunk or the trunk to the branches.

Tree diagram: a diagram used to determine the relationship between what needs to be accomplished and how to accomplish it. The sequence is always from the trunk to the branches.

TRIZ: Theory of Innovative Problem Solving, the result of nearly 50 years of research and development in the former Soviet Union by Ginrich Altshuller.

User: the individual or organization using a product or service.

Value engineering: an engineering discipline that finds less expensive designs to satisfy the original design intent.

Voice of the Customer Table (VOCT): the VOCT has two parts: Part one contains information on "who," "what," "where," "when," and "how" the product or service is or can be used. Part two sorts the reworded customer's voice into possible categories for analysis, such as reliability.

Bibliography

Akao, Yoji, ed., *Quality Function Deployment: Integrating Customer Requirements into Product Design*, translated by Glenn Mazur, 1990, Productivity Press: ISBN 0-91299-41-0.

Akao, Yoji, "History of Quality Function Deployment in Japan," *First European Conference of Quality Function Deployment*, March 25, 1992, Milan, Italy.

Altov, H. (pen name of Alshuller), *And Suddenly the Inventor Appeared*, 1994, Technical Innovation Center, Worchester, MA: ISBN 0-9640740-1-X.

Altshuller, Genrich S., *Creativity as an Exact Science*, 1988, Gordon and Breach, New York, NY.

Eureka, William E.; Ryan, Nancy E., *The Customer Driven Company: Managerial Perspectives on QFD*, 1988, ASI Press, Dearborn, MI: ISBN 0-941243-03-6.

GOAL/QPC Research Committee, *Quality Function Deployment: A Process of Translating Customers' Needs into a Better Product and Profit*, 1989, GOAL/QPC, Methuen, MA.

Griffin, Abbie, *Evaluating Development Processes: QFD as an Example*, Report No. 91-121 Marketing and Management, University of Chicago, Chicago, IL.

Hauser, John R.; Clausing, Don, "The House of Quality," *Harvard Business Review*, May-June 1988.

Hauser, John R., "How Puritan-Bennett Used the House of Quality," *Sloan Management Review*, Vol. 34, No. 3 1993.

Hollins W.; Pugh S., *Successful Product Design - What to Do and When*, 1990, Butterworth Ltd., London.

Imai, Masaaki, *Kaizen: The Key to Japan's Competitive Success*, 1986, Random House, NY: ISBN 0-13-952433-9.

Ishikawa, Kaoru, *What Is Total Quality Control?: The Japanese Way*, 1985, Prentice-Hall, Englewood Cliffs, NJ: ISBN 0-13-952433-9.

Kaelin, Oscar; Klein, Robert L., "How QFD Saved a Company - The Renaissance Spirometry System," *Transactions of the Fourth Symposium on Quality Function Deployment*, 1992, QFD Institute, Novi, MI (co-sponsored by ASI, ASQC and GOAL/QPC).

Kaplan, Stan, *An Introduction to TRIZ: The Russian Theory of Inventive Problem Solving*, 1996, Ideation International, Southfield, MI.

Kihara, Takami; Huchinson, Charles E., "QFD as a Structured Design Tool for Software Development," *Transactions of the Fourth Symposium on Quality Function Deployment*, 1992, QFD Institute, Novi, MI (co-sponsored by ASI, ASQC and GOAL/QPC).

King, Bob, *Better Designs in Half the Time: Implementing QFD Quality Function Deployment*, 1987, GOAL/QPC, Methuen, MA.

Klein, Robert L., "New Techniques for Listening to the Voice of the Customer," *Transactions of the Second Symposium on Quality Function Deployment*, 1990, QFD Institute, Novi, MI (co-sponsored by ASI, ASQC and GOAL/QPC).

Marsh, S.; Moran, J. W.; Nakui S.; Hoffher, G., *Facilitating and Training in Quality Function Deployment*, 1991, GOAL/QPC, Methuen, MA: ISBN 1-879364-18-2.

Mizuno, Shigeru, *Management for Quality Improvement: The 7 New QC Tools*, Productivity Press, Cambridge, MA: ISBN 0-915299-29-1.

Mizuno, Shigeru; Akao, Yoji, *Quality Function Deployment: Approach for Total Quality Control*, 1978, JUSE, Tokyo.

Mizuno, Shigeru; Akao, Yoji, *QFD: The Customer-Driven Approach to Quality Planning and Deployment*, 1994, Asian Productivity Organization, Tokyo.

Nakui, Satoshi "Cha", "Comprehensive QFD System," *Transactions of the Third Symposium on Quality Function Deployment*, 1991, QFD Institute, Novi, MI (co-sponsored by ASI and GOAL/QPC).

Nakui, Satoshi "Cha"; Terninko, John, "Structuring a Quality Design Process Chart," *GOAL/QPC 9th Annual Conference*, 1992, Boston, MA.

Ono, Michiteru; Ohfuji, Tadashi, *Expressing Demanded Quality on Quality Function Deployment*, 1990, Japanese Society for Quality Control, Tokyo.

Pugh, Stuart, "Concept Selection - A Method That Works," *International Conference on Engineering Design*, March 9-13, 1981, Rome, Italy.

Pugh, Stuart, *Total Design - Integrate Methods for Successful Product Engineering*, 1991, Addison-Wesley, Reading, MA: ISBN 0-201-41639-5.

Saaty, Thomas L., *Decision Making for Leaders: The Analytic Hierarchy Process for Decisions in a Complex World*, 1990, RWS Publications, Pittsburgh, PA: ISBN 0-9620317-0-4.

Saaty, Thomas L., *The Analytic Hierarchy Process: Planning, Priority Setting, Resource Allocation*, rev. and ed. 1990, RWS Publications, Pittsburgh, PA: ISBN 0-9620317-2-0.

Scheurell, Diane M., "Taking QFD Through to the Production Planning Matrix: Putting the Customer on the Line," *Transactions of the Fourth Symposium on Quality Function Deployment*, 1992, QFD Institute, Novi, MI (co-sponsored by ASQC, ASI and GOAL/QPC).

Scheurell, Diane M., "Concurrent Engineering and the Entire QFD Process: One Year After Start-Up of a New Mill," *Transactions of The Fifth Symposium on Quality Function Deployment*, 1993, QFD Institute, Novi, MI (co-sponsored by ASI and GOAL/QPC).

Sullivan, Lawrence, "Quality Function Deployment," *Quality Progress*, No. 6 (June 1986), ASQC, Milwaukee, WI.

Taguchi, Genichi, *Introduction to Quality Engineering*, 1983, Asian Productivity Organization, Tokyo: ISBN 92-833-1083-7.

Terninko, John, *Robust Design: Key Points for World Class Quality*, 1989, Responsible Management, Nottingham, NH.

Terninko, John, *QFD Needs Help*, 1990, ASQC Automotive Division, Detroit, MI.

Terninko, John, "Fanatic QFD User," *Transactions of the Second Symposium on Quality Function Deployment*, 1990, Novi, MI (co-sponsored by ASI, ASQC and GOAL/QPC).

Terninko, John, "QFD Synergy with Taguchi's Philosophy," *GOAL/QPC 8th Annual Conference*, 1991, Boston, MA.

Terninko, John, "QFD Assumes You Have an Imagination," *Transactions of the Third Symposium on Quality Function Deployment*, 1991, QFD Institute, Novi, MI (co-sponsored by ASI and GOAL/QPC).

Terninko, John, "Synergy of Taguchi's Philosophy with Next Generation QFD," *Transactions of the Fourth Symposium on Quality Function Deployment*, 1992, QFD Institute, Novi, MI (co-sponsored by ASI, ASQC and GOAL/QPC).

Terninko, John, "Does QFD Support a Corporation's 35-Year Vision," *Transactions of the Fifth Symposium on Quality Function Deployment*, 1993, QFD Institute, Novi, MI (co-sponsored by ASI and GOAL/QPC).

Terninko, John, "A Brief History of QFD and the Latest Improvements," *1st QFD Symposium*, 1994, Labein, Zamudio, Spain.

Terninko, John, "Efforts to Support and Encourage QFD in the USA," *1st QFD Symposium*, 1994, Labein, Zamudio, Spain.

Terninko, John, "Containment Ring Problem (Impeller Burst) Solved Using TRIZ," *Second Annual Total Product Development Symposium*, 1996, American Supplier Institute and Cal Poly University, Pamona, California.

Zlotin, Boris, unpublished articles, Ideation International, Southfield, MI.

Zultner, Richard E., "Before the House: The Voices of the Customers in QFD," *Transaction of the Third Symposium on QFD*, 1991, QFD Institute, Novi, MI (co-sponsored by ASI and GOAL/QPC).

Zultner, Richard; Terninko, John; Mazur Glenn, "A Recommended Curriculum," *GOAL/QPC 10th Annual Conference*, 1993, Boston, MA.

Zusman, Alla, unpublished articles, Ideation International, Southfield, MI.

INDEX

Worksheets

QFD Environment

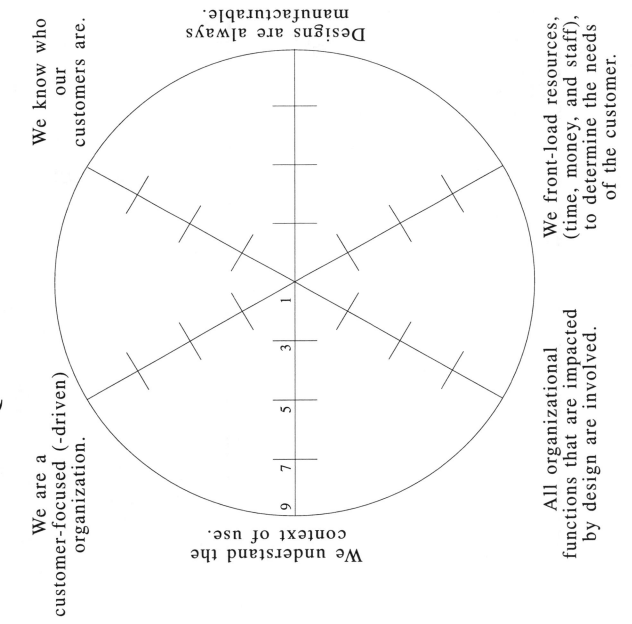

We know who our customers are.

Designs are always manufacturable.

We front-load resources, (time, money, and staff), to determine the needs of the customer.

We are a customer-focused (-driven) organization.

We understand the context of use.

All organizational functions that are impacted by design are involved.

1
3
5
7
9

INFO ABOUT PERSON	VOICE OF CUSTOMER

Worksheet A

I = inferred E = explicit	CONTEXT OF APPLICATION				
WHO	WHAT	WHERE	WHEN	WHY	HOW

Worksheet B

211

INTEGRATED DATA

Worksheet C

VOICE OF CUSTOMER TABLE

INFO ABOUT PERSON	VOICE OF CUSTOMER	I = inferred E = explicit	CONTEXT OF APPLICATION					INTEGRATED DATA
		WHO	WHAT	WHERE	WHEN	WHY	HOW	

For Sorting Customer Needs

	POSITIVE	NEGATIVE					EXPLANATION
		I REALLY LIKE IT	I LIKE IT	I FEEL NEUTRAL	I DO NOT LIKE IT	I REALLY DO NOT LIKE IT	1,C excitement need
		A	B	C	D	E	2,C not interested
1	I REALLY LIKE IT						3,C not interested
2	I LIKE IT						4,C minus evaluation
3	I FEEL NEUTRAL						1,D excitement need
4	I DO NOT LIKE IT						2,D not interested
5	I REALLY DO NOT LIKE IT						3,D not interested

EXPLANATION

1,C excitement need
2,C not interested
3,C not interested
4,C minus evaluation
1,D excitement need
2,D not interested
3,D not interested
1,E performance
        quality
2,E basic need
3,E basic need
4,E no effect

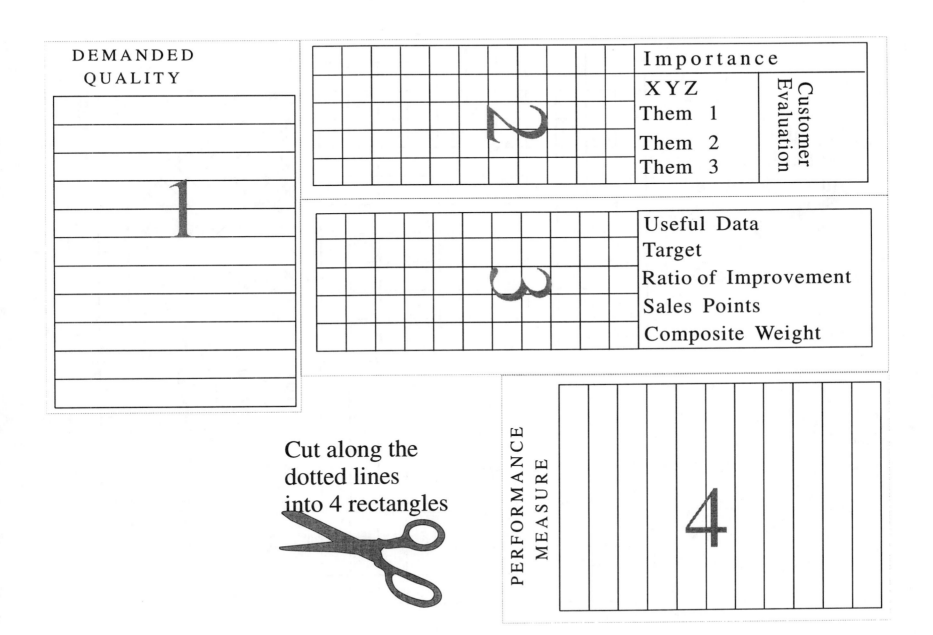

DEMANDED
QUALITY

1

Importance

X Y Z

Them 1

Them 2

Them 3

Customer Evaluation

Useful Data

Target

Ratio of Improvement

Sales Points

Composite Weight

2

3

Cut along the
dotted lines
into 4 rectangles

PERFORMANCE MEASURE

4

Cut along the
dotted lines
into 5 rectangles

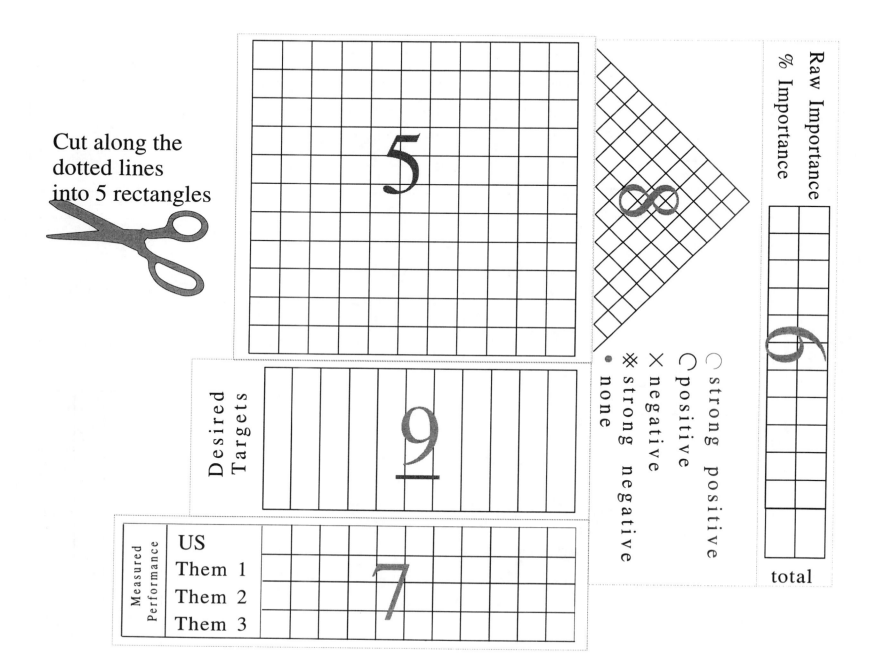

5

8

9

7

Desired
Targets

Measured
Performance

US
Them 1
Them 2
Them 3

◯ strong positive
◯ positive
✕ negative
※ strong negative
• none

Raw Importance
% Importance

6

total

216

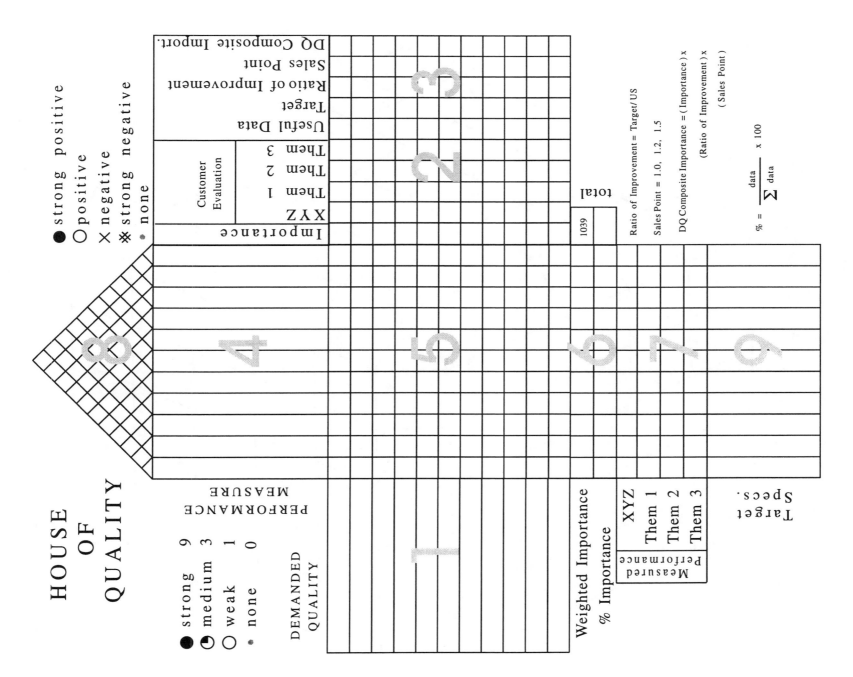

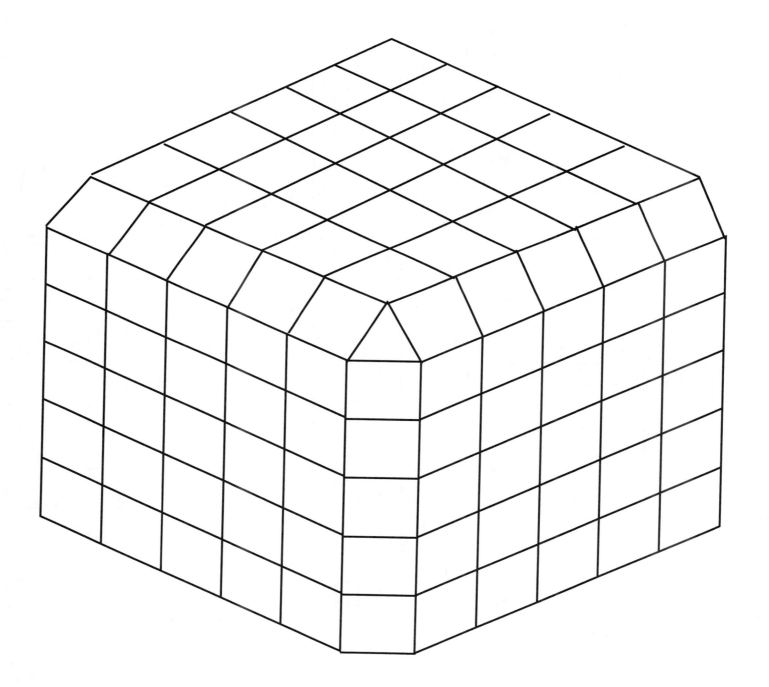

218

	CONCEPT	1	2	3	4	5	6	7	8	9	10	11	12
REQUIRE													
	+												
	−												

REQUIRE / CONCEPT	1	2	3	4	5	6	7	8	9	10	11	12
+ / −												

Acceptance
Matrix

Relationship

● Strong 9
◗ Medium 3
○ Weak 1
  None 0

Manufacturing Tests on Finished Product

Performance Measure		Importance	Specification					
Raw Importance								
% Importance								
Specification								

Manufacturing Worksheet

		Importance of Characteristic 1	Specification 2	Test Method Adequacy 3	Process Capability 4	Component Characteristic Process Component Steps Parameter	Input						Process				
							5	6	7	8	9	10	11	12	13	14	15
	A																
	B																
	C																
	D																
	E																
	F																
	G																
	H																
	Importance																
	Approved Operating Range																

Manufacturing Worksheet

	Importance of Characteristic 1	Specification 2	Test Method Adequacy 3	Process Capability 4	Component Characteristic Process Parameter Process Steps Component	Input 5	6	7	8	9	10	Process 11	12	13	14	15
A																
B																
C																
D																
E																
F																
G																
H																
Importance																
Approved Operating Range																